Galaxies

PAUL W. HODGE

Galaxies are among nature's most awe-inspiring and beautifully formed objects. In this highly informative and lucidly written book, Paul Hodge seeks to demystify galaxies and to examine closely our present-day knowledge of these magnificent star systems.

Hodge brings a historical perspective to his discussion of galactic research. He presents a summary of the revolutionary discoveries of the last decade, and he shows how they have contributed to our understanding of the nature and composition of the universe. Whereas previously perhaps a dozen astronomers devoted themselves to galaxy research, using two or three large telescopes, now hundreds of scientists are penetrating the mystery of the galactic world. This intensified research has yielded ground-breaking results: we are beginning to understand the enigmatic properties of the highly luminous yet relatively small quasars; we have a clearer understanding of the processes that generate spiral arms; we have a good idea of how different types of galaxies evolve; and we continue to grapple with the problem of the missing mass that is greater than anything detectable in the visible part of the galaxies.

This book succeeds in making the immense and remote universe of galaxies much more accessible to our imagination. It also conveys the excitement and wonder of this rapidly changing area of scientific inquiry. Enriched by numerous illustrations and written in an engaging style, *Galaxies* offers a nontechnical yet intelligent approach to the concepts and results of modern galactic research.

Paul W. Hodge is Professor of Astronomy, University of Washington.

THE HARVARD BOOKS ON ASTRONOMY
Edited by Owen Gingerich and Charles A. Whitney

Galaxies

GALAXIES

PAUL W. HODGE

HARVARD UNIVERSITY PRESS CAMBRIDGE, MASSACHUSETTS AND LONDON, ENGLAND 1986

Library of Congress Cataloging-in-Publication Data

Hodge, Paul W.
  Galaxies.

  (Harvard books on astronomy)
  Includes index.
  1. Galaxies. I. Title. II. Series.
QB857.H625 1986   523.1'12   85-8670
ISBN 0-674-34065-5

Preface

Nearly fifty years ago Harlow Shapley wrote a book for this series with the same title as this one. It became a classic. His enthusiasm, his gift for dramatic description, and his ability to arouse a reader's curiosity all made it an unusual example of the best in popular science. It was a readable and engaging book on an amazing topic, and went through three revisions, the most recent appearing in 1972.

Finally the flood of new facts and fresh problems about galaxies overwhelmed the book, and even Shapley's engaging style was not enough to keep it afloat any longer. A new vessel had to be launched. When the people at Harvard University Press asked me to launch it, I did not hesitate, though I certainly should have.

The task turned out to be as difficult as the prospect had been pleasing. The field of galactic astronomy had grown to such dimensions and complexity that a relaxed and informal book which would cover the subject seemed impossible. It still seems impossible. My attempt to navigate through current galactic astronomy misses much; so many new and interesting developments occur almost daily that it just is not possible to explore them all in one book. But if this introductory volume cannot satisfy the reader's curiosity about the magnificent objects we call galaxies, perhaps it can sustain and increase that interest. If so, the book will have served its purpose.

To acknowledge those to whom I am grateful for help on

this project, I will start with an impertinence: I want to thank Harlow Shapley for establishing a standard which I could not reach but which helped me avoid a number of trespasses against the science and the language. I also want to acknowledge the important assistance of Sandi Larsen, who typed the first chapters and then showed me how to set up the rest on the computer so that they would type themselves. Finally, I must thank the staffs of the observatories where most of the photographs used in this book were taken, many of them cheerfully taken by me: the Cerro Tololo Interamerican Observatory, the Harvard College Observatory, the Kitt Peak National Observatory, the Lick Observatory, the Manastash Ridge Observatory, and the Palomar Observatory. The men and women who have built and maintained these remarkable institutions are to be given credit for any inspiration that the reader may feel when viewing this book's photographs of some of nature's most glorious inventions, the galaxies.

Contents

Galaxies

Galaxies and
the Universe

For as far back as we can know, human beings have wondered what lies beyond the horizon and have set out to explore distant and uncharted lands. As the earth gradually revealed most of its secret places, astronomers began to open up an immense new realm of unmapped territory, beyond the confines of our tiny planet. Today, using modern telescopes and computers as their vehicles, explorers of the universe continue to push out to greater and greater distances, searching for the end of space, the ultimate frontier.

For centuries we were imprisoned in the solar system, believing the stars to be merely decorations on a sphere that lay just beyond the planets. Later, these pinpoints of light were recognized as distant suns, so far from us that their radiation takes years to reach the earth. Space seemed to be sparsely populated by these lonely stars, and scientists argued whether they extended off into the distance forever or whether there was an end to them, beyond which was emptiness. As astronomers explored farther into the depths of space, they did find an end. Our sun seemed to be one of an immense number of stars that resided together in a system they called the galaxy. Beyond the galaxy's edge was darkness.

The twentieth century brought about a new discovery: our galaxy is not the entire universe. Beyond the most distant stars in the Milky Way are other galaxies like ours, extending out to distances in space that are at the limits of our biggest telescopes. These magnificent star systems are one of modern astronomy's most active and exciting topics, and they are the subject of this book.

The Scale of Things

The Milky Way, a fairly typical galaxy of its type, is so immense that it takes light over 100,000 years to travel from one side to the other—at the speed of 300,000 kilometers per second, or about 670 million miles per hour. The earth and sun lie about 30,000 light-years from the center.* If we were to try to signal a hypothetical being who lived near the center of our galaxy, we would not get a reply until 60,000 years later. To put these distances another way, had we sent the message airmail on the day the universe began, at the airline speed of about 600 miles per hour, the message would only be about halfway there by now; and the reply would not get back before a total elapsed time of about 70,000,000,000 years.

Some galaxies are much bigger than ours. The largest—farflung galaxies which emit enormous energy in the form of radio waves, like the famous object in the southern sky called Centaurus A—have a total diameter a hundred times larger than the Milky Way's. On the other hand, many galaxies in the universe are comparatively small. Dwarf elliptical galaxies, like the tiny one in the constellation Draco, are only about 10,000 light-years across. Of course even these inconspicuous objects are almost unimaginably big. The Draco galaxy may be called a dwarf, but it is more than 100,000,000,000,000,000 miles from one side to the other.

Although there are billions of galaxies, they are not particularly crowded together. The universe is immense enough to accommodate them comfortably with plenty of space to spare. A typical distance between bright galaxies is about 5 or 10 million light-years, with little dwarfs taking up some of the space between. However, when their size is taken into consideration, galaxies are much closer to one another than are individual stars in our sun's neighborhood, for instance. A star's diameter is negligibly small compared with the distance to its nearest neighbor. Our sun's diameter is about a

---

* Two units of distance are frequently used in the discussion of galaxies—the light-year and the parsec. A *light-year* is the distance light travels in one year—roughly 6 million million miles. A *parsec* is a peculiar unit of distance that is defined in terms of the apparent movement (parallax) of stars across the heavens as the earth revolves around the sun. There are 3.26 light-years in a parsec, 1,000 parsecs in a *kiloparsec,* and 1,000,000 parsecs in a *megaparsec.*

million miles, whereas the distance to its nearest star is about 50 million times larger.

To imagine the huge distances between galaxies, think of galaxies scaled down to the size of people. In a typical part of the universe, then, the adults (the bright galaxies) would be separated by about 300 feet, with several small children scattered about between them. The universe would look like a somewhat expanded baseball game, with lots of open space between the players. Only in a few areas, where galaxies congregate in tight clusters, would our scaled-down universe be as crowded as a city sidewalk, and nowhere would it resemble a cocktail party or a rush-hour subway car. On the other hand, if the *stars* in a typical galaxy were scaled down to the size of people, the countryside would indeed be a lonely place. Typically, one's nearest neighbor would live about 60,000 miles away, or about 1/4 the distance to the moon.

These examples should make plain that galaxies are fairly sparsely spread out in the universe, and they contain mostly empty space. Even when we take into account the thin gas that lies between the stars within a galaxy, the density of matter is still extremely small. The universe of galaxies is huge, and it is very nearly empty.

Kinds of Galaxies

Galaxies do not all look alike. Some are smooth and round, some have a flattened, sprawling spiral form, and others show little discernible pattern. Astronomers (following the pioneering work of Edwin Hubble in the 1920s) divide the galaxies into three main categories, based on these shapes: elliptical galaxies, spiral galaxies, and irregular galaxies, abbreviated E, S, and Irr.

Elliptical galaxies are characterized by their overall elliptical shape and show no further structure other than a general decrease in brightness (luminosity) outward from a central point. The rate at which the brightness decreases outward follows a simple mathematical law, discovered by Hubble. To put this principle into the language of astronomy, we would say that elliptical galaxies have concentric elliptical *isophotes*—that is, if one draws a line connecting all the points where the luminosity of a galaxy appears to be the same, and then other lines for other luminosities (analogous to the lines of constant elevation on a topographic map), these lines will form a series of ellipses within ellipses, all with roughly the same center and roughly the same shape.

Subclasses of elliptical galaxies are designated by a number following the letter E. The number, n, is defined as

$$n = 10 \, (a - b)/a$$

where a and b are, respectively, the major and minor axes of any one of the galaxy's isophotes. Thus, an elliptical galaxy that has a circular outline will be classified as an E0, and a highly eccentric elliptical galaxy will be classified as E6 or so (figs. 1.1 and 1.2).

Spiral galaxies show a conspicuous flat, spiral arrangement of luminosity around a bulging nucleus. Ideal spirals have two arms emanating either from the nucleus itself or from a bar centered on the nucleus. These shapes have led to two principal subcategories, normal spirals (S) and barred spirals (SB). Normal spirals greatly outnumber barred spirals. Further subdivisions are determined by three criteria: (1) the relative size of the nuclear region, compared with the entire galaxy; (2) the tightness with which the arms are wound; and (3) the resolution of the arms into patchiness.

The Sa (and SBa) galaxies are those which have a very large nuclear region and very tightly wound spiral arms (nearly circular) that are continuous and smooth-appearing rather than patchy. The Sb (and SBb) galaxies have moderately small nuclear regions and more loosely wound spiral

1.1 NGC 1600, an E4 galaxy according to the Hubble scheme.

1.2 NGC 3377, an E6 galaxy. Very few elliptical galaxies are found that appear flatter than this one.

1.3 NGC 4594, a Hubble type Sa galaxy, with a large central bulge and diffuse, dusty arms.

1.4 NGC 3627, a Hubble type Sb galaxy, with brighter and looser arms than that in figure 1.3.

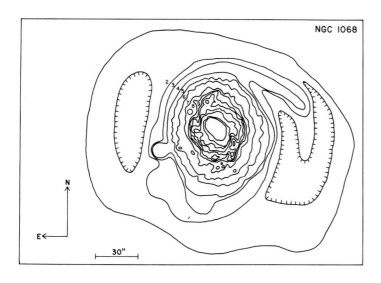

1.5 A contour map of the light distribution in NGC 1068, an Sb galaxy with an outer ring-like structure.

arms that are resolved to some extent into patches of light. Sc galaxies (and their barred counterparts) have very inconspicuous nuclear regions and very loose and fragmented spiral arms that are highly patchy (figs. 1.3 – 1.10). In SBc galaxies, even the bar is resolved into patchiness.

For all spiral galaxies, the nucleus is a luminous region with many of the characteristis of an elliptical galaxy. In fact, the luminosity law which Hubble found to apply to elliptical galaxies also applies to the central nuclear regions of the spiral galaxies, and so the nuclear region is sometimes called the "elliptical component."

Irregular galaxies make up the category into which Hubble placed all those that did not fit into the elliptical or spiral classes. Most irregular galaxies are rather similar to each other in appearance, being extremely patchy so that individual stars—particularly the many very bright ones—and regions of hot, luminous gas can be discerned. Some irregular galaxies have a conspicuous bar, and many of them have what appear to be patches of structure that suggest fragments of a spiral arm. Hubble realized that irregular galaxies of this type (called Irr I) seem to be extreme extensions of the spiral galaxy classification, where the arms are so patchy and disjointed that they cannot justifiably be described as spirals (fig. 1.11).

Other peculiar galaxies, lumped among Hubble's original class of irregulars, do not seem to be related to more common irregular objects, usually because of the presence of dust, distorted shapes, or other anomalous features (fig. 1.12). These were designated Irr II, but in subsequent revi-

1.6 NGC 1530, a barred spiral galaxy of
Hubble type SBb.

1.7 NGC 6946, a Hubble type Sc
galaxy, with wider, clumpy arms.

1.8 NGC 5457, also known as M101, a giant Sc spiral in Ursa Major. The arm pattern is not symmetrical, probably because of gravitational interaction with nearby galaxies. It is part of a small group of galaxies.

1.9 NGC 2403, a nearby galaxy of type Sc. It is near enough to us that we can resolve its brightest stars, including variable stars that give us its distance.

1.10 NGC 4303, a barred spiral of type
SBc. Dust lanes are noticeable on
opposite sides of the bar, inside the spiral
arm pattern.

1.11 NGC 1156, an irregular galaxy of Hubble type Irr I. This type of galaxy is usually inconspicuous and somewhat messy-looking compared with spiral galaxies.

1.12 NGC 3077, a dusty irregular galaxy of type Irr II.

1.13 NGC 1332, an S0 galaxy seen nearly edge-on. The shape is not perfectly elliptical, as in the E6 galaxy shown in Figure 1.2. The fainter galaxy seen above and to the left of NGC 1332 is a small companion galaxy of type E1.

1.14 NGC 2855, an S0 galaxy that shows faint outer dust lanes that make it look almost like an Sa galaxy.

sions of the Hubble classifications many of these strange-looking irregular galaxies were reclassified into other sub-classes. For example, galaxies with a flat disk like the spirals but no spiral arms are termed S0 galaxies (figs. 1.13 and 1.14). A few galaxies still remain outside the classification, and many of them have since been discovered to be either interacting pairs or subject to some other violent event.

Luminosity Classes

In 1960 Sidney van den Bergh found a series of morphological clues in the photographs of spiral galaxies that can be used to classify them according to their intrinsic luminosity. Whereas the Hubble classification scheme arrays the galaxies on a line in order of the degree of dominance and of contrast of the spiral structure, van den Bergh's luminosity criteria represent a plane perpendicular to that. A spiral galaxy of a given Hubble type, such as an Sc galaxy, which has bright, open, contrasting spiral arms, could be of any van den Bergh class from I to IV, the smaller number indicating the intrinsically brighter galaxy. Calibration by galaxies of known true brightness showed that the full range of luminosity classes represents a factor of about 5. Although it is admittedly qualitative, many astronomers have argued that tests of the van den Bergh classification indicate that it can be used, free of systematic effects, quantitatively as a measure of a galaxy's true, intrinsic luminosity.

Explaining Galaxy Types

Ever since Hubble, astronomers have tried to uncover the processes that give galaxies their shape. Some of the early theories attributed galactic shapes to an evolutionary progression. Galaxies were thought to start out as one type and gradually evolve into the other types. One idea was that they began as elliptical galaxies, developed spiral structure, and then disintegrated into irregular chaos. On the other hand, some astronomers thought it might work the opposite way: galaxies began in disorder, gradually wound themselves up as spirals, and ended their evolution with the pure symmetry and simplicity of ellipticals. Both theories were based on the premise that a galaxy's type reflects its age. Neither notion had any detailed physics to back it up; and after many years of research, both turned out to be wrong. Once astronomers had learned enough about the evolution of stars to be able to measure stellar ages (in the 1950s), they found that all types of galaxies are about the same age. Virtually all galaxies have at least some stars that are several billion years old, an indication that neither elliptical nor irregular galaxies are younger than spirals.

It *is* true, though, that elliptical galaxies are made up almost *exclusively* of old stars, while the other Hubble types contain relatively more young stars. Thus, the Hubble sequence of shapes does have something to do with ages. Apparently the rate of new star formation that has gone on since a galaxy's birth, and hence the distribution of stellar ages within a galaxy, is related to its shape. Elliptical galaxies have had little star-forming activity since their birth and so have few young stars, Sa galaxies are still forming stars but slowly, Sb's have more star formation going on, Sc's are rife with activity, while Irr I galaxies are the most active of all.

These facts have led to the idea that the Hubble sequence is a *conservation* sequence. The irregulars have conserved more of their gas and dust for star-forming, while the ellipticals used practically all of it up in one initial burst. But how did this difference lead to such different forms? This point will be taken up in Chapter 3, which summarizes our understanding of galactic evolution. Current ideas, now fully backed up with many kinds of evidence, indicate that the two most important things responsible for a galaxy's form are its initial conditions (mass and angular momentum) and its environment (whether it has close companions or is a member of a cluster). In this sense, galaxies are not unlike people: their personalities are shaped both by their heredity and by the company they have kept.

Galactic Structure

2

Galaxies are among nature's most beautifully formed objects. A photograph of a glistening spiral, taken with a modern giant telescope, is an awe-inspiring sight. Even when viewed through a small telescope, which can just barely capture a hint of a spiral galaxy's structure, these mysterious, faint images can be breathtaking.

Why do immense spiral arms wind around the centers of some galaxies? Why do others have bright bar-like features across their faces? How can we explain the smooth, structureless purity of yet others? The best way to answer these questions has been to measure the properties of galaxies' structure and to compare the results with mathematically constructed models of galaxies, built according to different assumptions about how galaxies were formed and what they contain. This chapter follows that plan; first we examine the various measurements of the structure of galaxies of various types and then we look at models to see how well they fit and to find out what we can about why galaxies look the way they do.

How to Build an Elliptical Galaxy

The simplest-looking galaxies are the ellipticals: smooth, uniform in color, and symmetrical. The near perfection of their structure hints at their essential simplicity, and it is certainly true that ellipticals have been easier to measure and to fit into theoretical models than have their more complex cousins.

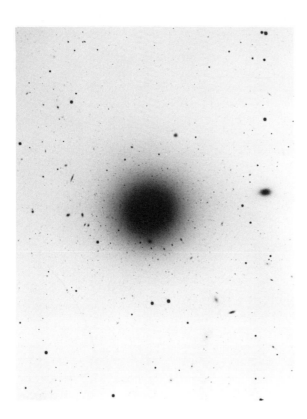

2.1 The elliptical galaxy NGC 1399, a giant E0 galaxy.

Let's look at the actual structure of a typical elliptical galaxy, NGC 1399, for example (figs. 2.1 and 2.2).* It has a bright nucleus at its center, which is surrounded by diffuse light that decreases steadily outward. As with all ellipticals, the rate at which the brightness goes down follows a simple mathematical expression. Also the outline of the galaxy — its shape — remains nearly the same at all light levels (fig. 2.3). Its isophotes are all, as near as can be told, perfect ellipses, centered precisely on the galaxy's nucleus. Their major axes line up in nearly the same direction and ratios of major to minor axes are all almost identical.

This basic simplicity is consistent with the idea that elliptical galaxies are presently governed by a very few forces. The stars' orbits are well mixed and smooth; nothing but gravity influences their arrangement, and little in the way of contin-

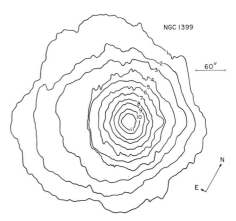

2.2 Isophotes (lines of constant brightness) of the large elliptical galaxy NGC 1399 in the constellation Fornax.

* Most galaxies are bright enough to have been included in the nineteenth-century catalog of nonstellar objects assembled by J. L. E. Dreyer and called the *New General Catalogue* (hence NGC numbers). Faint galaxies and radio galaxies are named after the (more recent) catalog in which they appear (for example, 3C 273, a quasar, is number 273 in the *Third Cambridge Catalog of Radio Sources*).

uous star formation has disrupted their regularity. When Hubble first pointed out most of these facts, he showed that the structure of an elliptical was very little different from that of a simple sphere of gas, arranged by gravity alone and full of identical particles of similar temperature. To build such a thing out of stars, you only need to take a large number of similar stars, place them near one another in space, let gravity go to work on them, and wait a long, long time until their motions all become similar. You should not give them any global systematic motions, such as rotation, and you must make sure that you have chosen quiet, well-mannered stars that will not erupt, eject matter, or otherwise disturb the boring monotony of unchanged stardom. But it is not necessary to arrange them in a perfect sphere at first. You can let them out of a rectangular box, for example, and just wait awhile; they will arrange themselves in a spheroid eventually. Gravity acts with spherical symmetry and if your galaxy has only gravity governing it, it will smooth out, lose its rough edges, and become a nice elliptical galaxy.

Real elliptical galaxies are not perfect spheres, of course. The isophotes of NGC 1399, for example, are ellipses rather than circles, and their axial ratios vary slightly at different distances from the center, making the isophotes less circular in the outer parts. Their alignment also twists a little. All these imperfections tell us that the simple model of elliptical galaxies is not quite right. Some history and special circumstances must have influenced the stellar orbits in a perceptible way. Perhaps some rotational motion is involved, perhaps tidal action from neighboring galaxies has been a factor, or maybe we are seeing the effects of special initial conditions so strong that gravity has not had time to wipe them out completely.

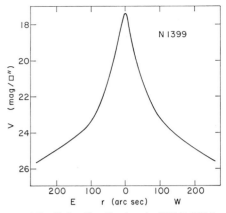

2.3 The light distribution in NGC 1399, showing a typical luminosity profile for an elliptical galaxy. The vertical axis gives the brightness (in magnitudes) of points along an east–west line through the center of the galaxy, measured per square arc second.

Disks and Bulges

Unlike ellipticals, spiral galaxies are characterized by disks and bulges. The arms are minor in terms of the number of stars they contain though they are important, conspicuous features. (In the same sense, the eyes of the human face are a small part of the body, but they command our attention and reveal a great deal about what is going on inside.)

The disk of a spiral galaxy is quite flat. Those spirals that we happen to see edge-on tell us that the thickness of a typical disk is about 1/10 its diameter. For our own galaxy, where we can count stars in the disk and measure its thickness, we find that the stars thin out rapidly, being quite sparse at about 3,000 light-years above the plane. This is especially true for the youngest stars and for the raw materials (gas and dust) that lie waiting to form future stars. Some edge-on spirals show dramatic lanes of dust across the very middle of the disk that are pencil-thin, whereas the oldest stars in a disk form a much thicker band.

Measures of the light in spiral galaxies' disks show an important commonality that is well documented but not satisfactorily explained yet. The brightness decreases outward in a fairly regular and universal mathematical pattern, but one which is different from that of elliptical galaxies (fig. 2.4). All disks follow this pattern, from those of tiny dwarf galaxies like GR8 to those of supergiant spirals like M101.

Computer-generated models of rapidly rotating systems of stars make what we see of galaxies' disks seem quite natural. Consider the elliptical galaxy that was described above. If its pre-galaxy cloud of gas could be set to spinning rapidly before most of the stars were formed, it would flatten out and the eventual distribution of stars would resemble the disk of a spiral galaxy. Thus, it appears that the essential structural difference between elliptical galaxies and spirals is the amount of initial rotation.

Then where does the bulge come from? If a rapidly rotating pre-galaxy cloud produces a disk, while a slowly or nonrotating one produces an elliptical galaxy, what are those fat, ellipsoidal bulges doing at the centers of spirals? They have most of the structural properties of elliptical galaxies: regular isophotes, old stars, considerable thickness, and smoothly decreasing light distribution. The answer seems to be found in the important fact that gas behaves very differently from stars. A cloud of gas can get rid of energy fairly easily by simply heating up and radiating the energy away. As it does so, a rotating gas cloud will rapidly flatten to a disk. However, if at any time the gas starts to condense into stars, the situation will change. Stars do not collide as atoms of gas do. They are much too small relative to the distances between them. Since they do not heat up as a result of collisions, stars do not dissipate energy efficiently, and hence they do not collapse to a plane. If, therefore, stars start to form — and they will do so first in the central regions where the densities are greatest — then they will stay where they are, in a big thick central bulge.

In the Milky Way, for instance, the first stars to form, which are now the oldest, are in this central bulge. The remaining gas eventually collapsed to the plane, where slowly more stars formed and rotated around the center

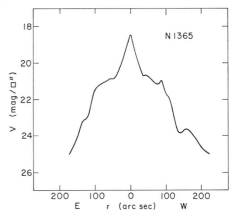

2.4 The distribution of light in the SBc spiral galaxy NGC 1365.

along with the gas. This thin, flat disk became the seat of most subsequent activity in our galaxy; stars, giant molecular clouds, excited gas clouds, and large-scale spiral patterns all evolved there, in an intricate structure that presently challenges our theoretical models.

The Riddle of Spiral Structure

Spiral galaxies would not look very interesting without their spiral structure; they certainly would not be spiral galaxies, but the point is a little more subtle than that. If a spiral galaxy forms because rotation forces its gas to collapse to a plane, then the spiral shape of the arms is seemingly a natural result, much like the pattern of cream stirred into a cup of coffee, or storms over the Caribbean, or water going down the drain. These are not strictly analogous to the case of a galaxy, but the connection is illustrative: where there is rotation there is likely to be spiral structure. For many years, therefore, astronomers were not particularly concerned with the fact that many galaxies had a spiral shape — it seemed quite natural.

The first serious difficulty arose when someone thought to ask the question, how long can a spiral arm last in a galaxy? We know the rotation periods of galaxies, which are typically a few hundreds of millions of years for stars that lie at about the sun's equivalent distance from the nucleus. We know the ages of nearby galaxies, which all appear to be about 10 or so billion years. If the spiral structure results from the fact that the inner part of the galaxy is rotating at a different speed from the outer part, then the arms should gradually wind up in a spiral pattern. But the number of windings should be very large for a galaxy as old as those around us, about as many as the age divided by the mean period of rotation, which would be 100 or so. Actual spiral galaxies, at least those that have clear, continuous spiral arms, only show one or two windings. Thus, the puzzle. Do spiral arms "freeze" somehow, so that they can persist? Or do they keep winding up to oblivion, to be succeeded by new ones? Or is there some way whereby they do not partake in the general rotation of the stars and gas, so that they rotate more slowly?

The problem is not that we cannot think of a way to make spiral structure — any blob that rotates like a galaxy, with different rotation periods at different distances, will make a spiral pattern. The problem is to find out how a galaxy makes a spiral shape that can persist. Currently, there are three different kinds of answers, and we are not sure yet which is right. It may be that all are right in one or another case and that the spiral structure in even one individual galaxy may have a mixed parentage.

Perhaps the neatest and most elegant explanation of spiral galaxies is known as the density wave theory. After various related theoretical ideas had been developed by the Swedish astronomer Bertil Lindblad, the density wave theory was fully worked out and applied successfully to galaxies by C. C. Lin and his students at the Massachusetts Institute of Technology in the 1960s. They showed, through mathematical analysis of the stability of a flat disk of stars, that an irregularity in the initial distribution of gas can become stable and will gradually form into a two-armed spiral pattern that rotates much more slowly than the stars. As stars move into the arm, they slow down for a while, giving the arm its high stellar density, and then they move on out the front of the wave. At the front edge there should be a shock wave in the gas, which could trigger star formation, thereby explaining why, in some real galaxies, one finds a concentration of active gas clouds and newly formed stars in the arms (fig. 2.5). The shapes of the spiral arms in this hypothesis are very much like the shapes of real spiral arms in a small subset of galaxies that have "perfect" spiral structure, such as M81. But they do not fit the more common kind of galaxy, which has highly imperfect arms—fragmented, vague, and indistinct.

The theory that works best for the more informal galaxies builds on the rather simple distortions in any structure that naturally result from a galaxy's rotation. Instead of a persistent set of arms, this idea predicts the continuous birth and death of spiral segments. Many pioneers in the field realized that this method might work; they merely needed to find a way of regenerating the arms. By 1965 a computer movie was made that showed the entire process in action, using the

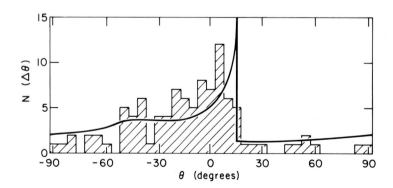

2.5 The distribution of HII regions in a galaxy arm can be plotted to show the density of star-forming regions, going across the arm from the inner edge to the outer. This figure shows a comparison of the spiral arm structure of NGC 3136 (which is represented by the striped histogram) with the density wave theory predictions (which are represented by the solid line). The fact that the two curves look similar suggests that the arms of this galaxy are primarily governed by a density wave.

galaxy M31 as a model and adopting a random (stochastic) pattern for the emergence of star-forming regions. As these regions are born, they show up as bright patches of activity; differential rotation then draws them out into long, narrow segments of a spiral shape, and they gradually fade as the gas concentrated in them is used up. The result is not a perfect two-armed spiral pattern, of course, but rather a collection of spiral segments that cover a galaxy and that give it a certain spiral shape, but with arms that cannot be traced around the center more than a few tens of degrees.

The shapes produced by the computer movie actually resemble many spiral galaxies, and so it is likely that a stochastic process of this sort dominates in such objects. This is particularly true for certain kinds of ideal star-forming regions, which have a sequence of areas in various stages of activity; at the front there is a giant molecular cloud about to condense into a star cluster, behind it is a gas cloud illuminated and cleared out somewhat by the presence of stars just formed, and behind that is a star cluster which is relatively clear of gas and is aging and slowly dispersing. This sequence of regions is roughly linear and will be pulled out into a spiral arm segment by differential rotation. The result is a spiral galaxy made up of disjointed spiral arm segments. The stochastic theory, therefore, seems able to explain just those galaxies that fail to be accounted for by the density wave model. Thus, we may need no more ideas, just the patience to obtain the detailed measurements necessary to check the properties of spiral arms against various versions of each theory.

There is, however, another possibility. Any disturbance of the disk can lead to a bunching up of the gas in a way that will show up as spiral arms or spiral segments. The disturbance might come from either outside the galaxy or from inside its nucleus. Of the former possibilities there is the chance that interstellar gas might stream into the galaxy, forming arms as it does so. This idea is not very attractive because the gas would preferentially come in at the poles, where there isn't already gas to collide with, and we know of very few cases of spiral arms that are not in the plane of the disk. A more attractive outside agent would be the tidal action of other galaxies during close encounters. The tides raised by a near collision will involve stars and gas and could disturb the galaxy's shape enough to lead to an irregular pattern that will rotate into a spiral shape. This is a nice idea, but it has the disadvantage of requiring a close encounter with another galaxy. Unfortunately, galaxies' separations are too large for the mechanism to be very effective in most cases. We may still have some surprises, however, on the subject of galactic encounters. Recent measurements of star-formation rates show that galaxies very near each other seem to have an abnormally high level of star formation going on, especially in their nuclei. Perhaps it will turn out that the tidal effects are more easily triggered than we now think.

There is no compelling evidence to suggest that spiral arms might result from action in the nuclei of galaxies, but enough things go on in those mysterious and violent places to make the idea come up. Radio galaxies (Chapter 11) and quasars (Chapter 12) all involve highly energetic events in galactic nuclei, many of which eject immense streams of gas out even beyond the visible galaxy. It may be that this kind of activity could somehow lead to spiral arms; at present, though, this idea is vague and unencumbered by any reasonable physical model.

Bars

Many spiral galaxies have another remarkable structural feature, usually connected in some way to the spiral arms: a large bar-like concentration of stars that crosses the nucleus and extends symmetrically on each side (figs. 2.6 and 2.7). Their measured velocities indicate that these bars tend to rotate around the nuclei like solid bodies, though of course they are actually individual stars and gas. Bars that occur in S0 or Sa galaxies are smooth and made up entirely of stars, while those in Sb, Sc, or Irr galaxies often have lots of gas and dust in them as well. Arguments still rage about the motions of the gas in these bars; some evidence shows it flowing outward along the bar and other data indicate that it flows in. In any case, the existence of a bar is not so very surprising to astronomers who study the dynamics of galaxies. Numerical models indicate that instabilities in the disk of a rotating galaxy can show up in the form of a bar that resembles those we have observed.

Finding the Order in Chaos

Irregular galaxies are not completely irregular in their characteristics. They have a few features in common with each other that serve to hint at the reasons for their apparently chaotic shapes (fig. 2.8). They are all rich in gas, and almost all of them have lots of young stars and clouds of glowing, ionized gas, often exceptionally large and brilliant. None has a central bulge or any real nucleus. The light of irregular

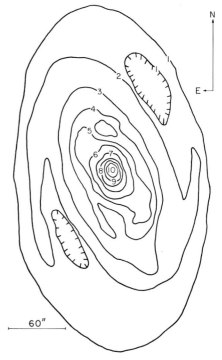

2.6 Isophotes of the SBa galaxy NGC 1350, showing the bar-like structure in its central regions.

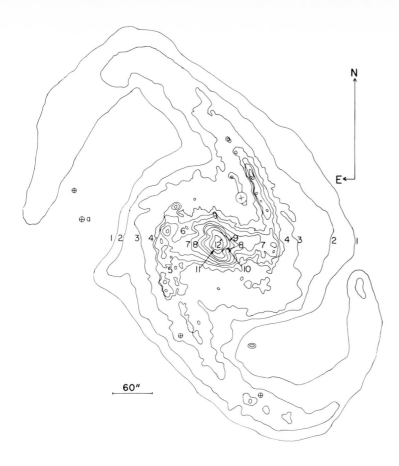

2.7 Isophotes of the SBb barred spiral galaxy NGC 1365.

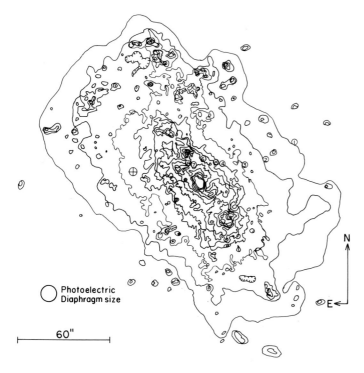

2.8 Isophotes of the irregular galaxy NGC 4449.

galaxies, on the average, decreases in intensity outward from the center according to the same mathematical law as in spiral galaxies. Many of them have bar-like structures in their central areas, the Large Magellanic Cloud being a particularly good example (see Chapter 6).

An important hint about how irregular galaxies form comes from a comparison of their total luminosities with those of spiral galaxies.* They are almost all much fainter than even the brightest spiral galaxy. The spiral M33, which represents about the lower limit to the luminosity range for spirals, is still brighter than the Large Magellanic Cloud, which is among the brightest known irregulars. The lack of spiral arms in irregular galaxies, therefore, seems to be related to their smallness. Probably it also relates to the amount of angular momentum and the amount of turbulent motions within them. The planes of irregular galaxies are relatively thicker than those of spirals, suggesting a rotation of stars and gas slow enough to keep spiral arms from forming. Were the rotation extremely small, on the other hand, the galaxy would not have collapsed to a plane at all, thick or thin, and a low-mass dwarf elliptical galaxy might have formed.

We are not really sure of the relationship between dwarf elliptical galaxies and dwarf irregulars. The traditional view is that the stars in ellipticals are very old (10 or more billions of years old), while the irregulars have both old and young stars. However, there is some evidence to suggest that some dwarf ellipticals—the Carina dwarf, for example—were still actively forming stars only 2 or 3 billion years ago, and

* The brightness or luminosity of a star or galaxy is described as its *magnitude.* Stars with small-number magnitudes are bright and those with large-number magnitudes are faint. A first magnitude star is among the dozen or so very brightest stars in the night sky, while a sixth magnitude star is about as faint a star as can be seen without a telescope. The faintest galaxies that have been detected are about 24th magnitude. This scale, which is logarithmic, is set up so that each group of 6 magnitudes corresponds to a factor of 100 in brightness.

Usually magnitudes that are quoted for objects refer to their *apparent* brightness as seen from the earth. But when astronomers wish to describe the *intrinsic* brightness of a star or galaxy, they refer to the *absolute magnitude,* which is the magnitude an object would have if placed at a standard distance of 10 parsecs. The absolute magnitude of the Andromeda galaxy, for example, is −21.6—extremely bright—while its apparent magnitude is only 4.4. The great difference is the result of its being so far away from the earth, about 600,000 parsecs (2,000,000 light-years) instead of the standard 10 parsecs used to determine absolute magnitude.

during those episodes they may have looked like dwarf irregulars. This is an important issue, because the dynamical explanations of the differences between the two will have to be discarded if it is found that they freely change from one to another and back again.

There is still much to learn about the structure of galaxies, though we are making progress. We can do more than just describe the differences; we can point to explanations for many of them. But the number of unsolved problems is substantial enough to keep astronomers thinking creatively for many more years.

The Formation and Evolution of Galaxies

3

One of the goals of modern astronomy is to understand how galaxies were formed and how they evolve. In the days of Edwin Hubble and Harlow Shapley, it was tempting to believe that the classes of galaxies might correspond to different stages in their development. But that idea turned out to be wrong, and the task of reconstructing the life histories of galaxies turned out to be difficult. Most puzzling has been the problem of how galaxies came into being in the first place.

The nature of the universe before galaxies existed is unknown, and the hypothetical characteristics one attributes to it depend to a great degree on the cosmological model one chooses. Most currently accepted cosmological models posit a general expansion from time-equals-zero, shortly after which the density and temperature of the universe are both extremely high. The physics that describe the primordial explosion in these models can be followed fairly reliably right up to the moment when density and temperature were low enough to permit galaxy formation. It took about 1 million years for things to cool off and expand enough for matter to become important in the universe. Prior to that time, radiation (intense light) predominated and chunks of matter such as stars or galaxies could not have formed. But when the temperature was about 3,000K and the density was about 10^{-21} gm/cm^3 (quite a bit less than that of Earth's atmosphere but greater by at least a billion than the overall density of the present universe), matter could form at last. At that time only the atoms of hydrogen and helium could have been produced in significant amounts out of the cosmic cloud.

Though one can imagine several mechanisms for forming galaxies out of this hydrogen and helium gas, finding even one model that will actually work under the probable conditions in the early universe is difficult. For a generally expanding, smoothly distributed universe of uniform temperature, there is very little reason for a galaxy to form; such an idealized universe will never have a galaxy in it. The existence of galaxies in the universe and their apparent dominance as forms of matter argues that the pre-galaxy environment was not anything like such an idealized gaseous cloud. Instead, there had to have been irregularities of one sort or another. But what kinds of irregularities, and where did they originate?

Instability

Most attempts to find ways by which the material in the universe could have condensed into galaxies are based on an idea first worked out in detail by Sir James Jeans. Though it is now assumed that the early universal gas expanded according to a relativistic cosmological model, Jeans' ideas were based on a simpler, Newtonian universe, in which gravitational instability occurs when a pocket of denser material (called a perturbation) becomes sufficiently dense and sufficiently small. The characteristic size of the density pertubation that is just barely unstable is called the *Jeans length* and is found to depend on the speed of sound in the medium, the constant of gravity, and the density of the material.

The *Jeans mass* is defined as the mass of material that would begin to be unstable and would begin to contract under its own gravitational field (fig. 3.1). At the beginning of the "era of matter," the Jeans mass is calculated to be about 10^5 solar masses; so at that point in the history of the universe, perturbations with masses as large as this value, which includes all known galaxies, would have been unstable and would have contracted. It is not possible with the simple Jeans model to examine the situation during the "era of radiation" because the effects of radiation pressure on gas during those times are not included in this simple analysis. However, several astronomers and cosmologists have examined the more complicated situation with radiation present, and the results roughly agree with the simpler models.

In search for a type of irregularity or instability that will

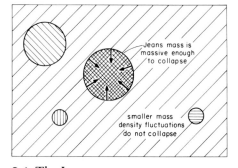

3.1 The Jeans mass.

lead to the present universe of galaxies, astronomers have examined many types besides gravitational ones. Among them are possible imbalances in the ratio of matter to antimatter, thermal instabilities, fluctuations due to ionization and its dependence on temperature, and variations in the distribution of charge.

If it is hypothesized that, for the sake of symmetry, the amount of matter in the universe is and was equal to the amount of antimatter, then the present existence of matter and antimatter in isolated pockets of the universe might naturally be the result of slight local inequalities in the early universe, after matter and antimatter separated from radiation. During expansion, complete annihilation would take place in those areas where matter and antimatter are equal, whereas in areas with an initial excess of one or the other, some matter or antimatter would remain (fig. 3.2). The distribution of matter and antimatter would be lumpy, and the lumps would contract to form clusters of galaxies. Such a universe would end up with bits and pieces of matter and antimatter in different places. About half of the galaxies that we see, in such a case, would be made up of antistars. Were we to travel to such a place and try to land on an antimatter planet, our atoms would combine violently with the antimatter atoms of the landing site to annihilate each other, producing a bright flash of light but not a very pleasant visit. There would be nothing left of us and only a hole in the ground to commemorate our adventure.

A more likely hypothesis would be that the amount of matter was just slighter greater than the amount of antimatter at the start. Then most of the matter would annihilate all of the antimatter in the early high-density cosmic phases, leaving a universe bathed in light and laced with just enough matter to form the galaxies.

Another mechanism that might have contributed to the condensation of matter is thermal instability. Regions of slightly higher density cool off more rapidly than their surroundings. The hotter regions around them compress them even further, increasing their density. Thus, a small density perturbation can become increasingly more unstable (fig. 3.3).

Yet another idea, first suggested by George Gamow, is that gravitational forces can be amplified by a "mock gravitation" produced by the bright field of radiation in the early history of the universe. Particles in such a universe tend to shadow one another from radiation and, as a result, experience a force toward each other. This force felt by particles

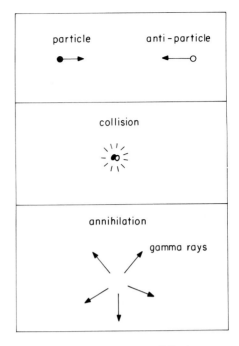

3.2 Matter–antimatter annihilation.

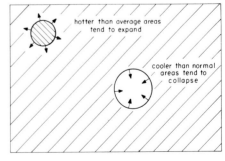

3.3 Temperature fluctuations.

toward one another has an inverse square behavior like gravitation. One can imagine, as an example, two particles separated by a small distance in a radiation-rich field. Particles absorb energy from the photons in the field and are thereby experiencing forces in various directions. Consider the situation when one particle absorbs a photon coming from the direction opposite to the direction of the second particle. That particle experiences a force toward the second particle. Because that photon has been absorbed by the first particle, the second particle is protected from the radiation field in that direction and therefore experiences a preferential force toward the first particle. The result has the effect of a mutual attraction between the two particles due to the shadowing of each other from the radiation field. It is found that this shadowing effect is important only in the first 100 or so years of the universe, after which the intensity of light and proximity of particles diminishes.

Contraction

After individual protogalaxies have achieved their gravitational identity through some form of instability in the pregalactic gas, they collapse to form galaxies of much smaller dimensions and higher densities, leaving the intervening space virtually empty. The actual means by which the contraction takes place is something that we can explore only by theoretical modeling. We have not yet discovered a galaxy that we can say for sure is very young compared with the estimated age of the universe, and there is nothing, then, that can be observed in the contraction stage. Instead, we must explore those clues about the pre-contraction environment that can be extracted from the present characteristics of galaxies and from what we can learn of the past by looking far into the distance. We can also approach the problem by proposing plausible initial conditions and carrying through the calculations to see whether we can arrive at a realistic picture of a contracted galaxy. The initial conditions that we must begin with for these calculations include the mass of the galaxy, its angular momentum, its dimensions, its temperature, its chemical properties, its magnetic field, and the turbulent motions within it.

Consider the simplest initial situation, in which a protogalaxy has properties such that it is cold, uniform in density throughout, perfectly spherical, and without turbulence, rotation, magnetic field, or external influence. For an object with a mass comparable to that of the Milky Way — on the order of 10^{11} solar masses—such a set of initial conditions leads to a completely unbraked collapse. The gravitational potential of such an object is sufficiently large that no physical process can stop its collapse into a massive black hole, and calculations indicate that in a cosmically short time such an object will disappear (fig. 3.4). It passes through the *Schwarzschild limit*, which is the general relativistic limit that occurs when a massive body contracts to such a small size and enormous density that light can no longer escape from it. It disappears to an outside viewer, and only its gravitational field is detectable. Thus, the very simplest initial conditions would not lead to a galaxy at all.

A more reasonable set of initial conditions would be the following: By one of the processes discussed above, the gas cloud has already contracted to the point where the density is high enough for it to be stable in spite of the expansion of the surrounding universe, let us say approximately 10^{-28} gm cm^{-3}. Assuming a mass of 10^{11} solar masses, then this density for a spherical cloud implies an initial radius of about 200 kpc (as opposed to 30 kpc, which would be typical for a galaxy of this mass *after* contraction). For contraction to occur, the kinetic, magnetic, and gravitational energy must all be appropriately balanced. The other initial conditions necessary for contraction to begin are that the rotational velocity must be small, less than about 40 km/sec, any turbulent velocity must be less than around 30 km/sec, the temperature must be less than about 2×10^5K and the strength of the magnetic field must be reasonably small, less than about 2×10^{-7} gauss.

If the cloud remains uniform in density distribution during contraction, then the gravitational energy rises in inverse proportion to the shrinking radius. The temperature, on the other hand, remains approximately the same until the density of the material is so large that it becomes optically thick to the wavelengths of light emitted. Before this happens, the thermal energy (the amount of energy represented by the motions of the gas particles—that is, the temperature) of the gas cloud does not depend on the radius, but after the density reaches this critical value, the thermal energy rises dramatically as the radius decreases. The thermal energy of the cloud can stop the contraction only when the radius is less than this critical value, the thermal limit. The turbulent energy is not important when the cloud is larger, because it is rapidly dissipated.

Similarly, the magnetic energy, which increases as the cloud shrinks, never becomes larger than the gravitational

A non-rotating protogalactic cloud

collapses,

becoming a black hole

3.4 The fate of a non-rotating protogalaxy.

energy if it started out smaller to begin with. Eventually, the radius becomes small enough that the rotational energy balances the gravitational energy; this defines the *rotational limit.* At yet another critical size, stars condense out of the gas and there begins a rapid change from a gas cloud to a galaxy of stars. This is the *condensation limit.* The ultimate fate of the contracting cloud depends on the relative sizes of these three critical radii. Three interesting possibilities emerge, depending on which is largest.

When the rotational limit is the largest, contraction is stopped by the rotation (fig. 3.5). The centrifugal forces, however, are limited to the plane of rotation in such a way that contraction continues perpendicular to this plane until a thin disk is formed. This thin disk is conspicuous by its shape and by its rotation; it is a spiral galaxy.

In the case where the condensation limit is the largest, then star formation begins before the effects of the rotation become important braking factors in the contraction. As the density increases, the rate of star formation becomes greater and most or all of the gas goes into star formation. Then, when the contraction is finally halted at the limit, there is little or no gas left to dissipate energy efficiently. Therefore, a disk does not form. Energy requirements say that the object must then expand somewhat until the radius reaches another critical value. The stars will orbit in such a way that the galaxy will be nearly spherical, depending on the actual amount and distribution of the initial angular momentum. With these properties — the nearly spherical shape, the lack of gas, and the large number of stars formed near the beginning of its history — the object is clearly going to be an elliptical galaxy (fig 3.6).

In the third case, when neither the rotational limit nor condensation limit is large enough to brake the contraction, the cloud gets smaller and smaller, until finally a supermassive star-like object is formed. This will probably be a black hole, invisible and almost undetectable.

Watching Galaxies Evolve

Once a galaxy has taken shape, the next stages are slow and far less spectacular. Stars form, they die, and they eject material rich in heavy elements which forms new stars. The galaxy gradually dims and reddens, and the chemical composition of its population of stars slowly changes as the heavy elements are enriched in the gas and dust from which succeeding generations of stars form.

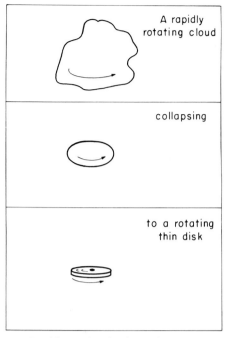

3.5 Rapid rotation leads to the formation of a flat plane.

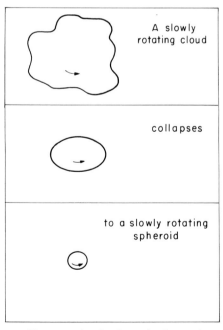

3.6 Slow rotation leads to the formation of an elliptical galaxy.

We cannot see a galaxy change. A human life is at least 1 million times too short. But we can see the effects of evolution by looking back to earlier and earlier stages in the evolution of our universe, when the galaxies all appear younger. The very most distant normal galaxies that we can now see are younger than our neighbors, as we see them. Light from a galaxy that is 10 billion light-years away, for instance, has taken 10 billion years to reach us, so the image that we see and measure is that of a 10 billion-year-younger galaxy. If the universe is 15 to 20 billion years old (we are not yet sure of the exact figure), then the galaxy as observed would be only about one-third the age of the galaxies near us, where the travel-time of light is short. Of course, this argument depends upon the belief, which is supported by local galaxies and predicted from cosmological models, that all galaxies contracted and formed at about the same time, shortly after the Big Bang.

To see galactic evolution, we need merely to look increasingly farther into the distance. The first couple of billion light-years is too small a distance to detect change, but the galaxies beyond that show real differences, especially noticeable in their colors. Recently, a truly evolutionary effect in the colors of galaxies was observed at around 10 billion light-years away. Using special detectors with the Palomar 200-inch telescope, astronomers have looked at 23rd and 24th magnitude galaxies with sufficient accuracy to see what young galaxies look like. Much as the theoretical models predict, galaxies were both brighter and bluer back then.

The calculations of the Yale astronomer Beatrice Tinsley, who devoted much of her brief but creative life to the study of galactic evolution, have helped astronomers to understand the details of these aging effects. From models produced by Tinsley and her co-workers, we know that the rate of dimming and the color changes depend on many things: the distribution of mass among the stars, the rate of regeneration of material into stars, the percentage of stars formed in any initial burst, and many other considerations. Currently observed distant galaxies are beginning to give us these details. It is an amazing thing to be able to learn about happenings that take billions of years to occur. We do it by turning the clock back to billions of years ago, looking billions of light-years into the distance.

Another conspicuous difference between the young galaxies in the distant universe and galaxies nearer the present is that there were many more active or explosive galaxies back then. Both quasars and radio galaxies increase in density as we look farther and farther away. Therefore, these things must have been more common in the early years of the universe. Current theoretical models suggest that they are generated by the collapse of supermassive objects, perhaps black holes, at the centers of galaxies. Black holes are harmless enough if there is nothing to drop into them, but become agents of violent energy if gas or stars come too close to their gravitational field. Probably young galaxies, still full of unprocessed gas, were much more likely to be feeding this gas into their central cores than do old galaxies today. If black holes were lurking there, these galaxies would be more likely to flare up as quasars or radio galaxies (see Chapters 11 and 12). Now, apparently, most such extremely violent activity is over.

Stars and Clusters, Gas and Dust

4

Galaxies are made of many things. They can contain stars, star clusters, glowing gas clouds, and dust. Invisible except to radio telescopes are neutral hydrogen gas, molecules, and clouds of fast, charged particles. Invisible to every means of detection, so far, are the unknown objects that make up, perhaps, the bulk of a galaxy's mass, those mysterious, pervasive things that occupy some kind of massive halo (Chapter 5). We know the most, of course, about the visible things in a galaxy, and they are capable of telling us a great deal about the details of a galaxy's life story.

Stellar Populations

One of Walter Baade's many contributions to the understanding of galaxies was his division of stars into two populations. Using the Mt. Wilson 100-inch telescope, he had explored nearby galaxies, especially the spiral M31 and its companions, and compared them with stars and star clusters in the Milky Way Galaxy. During the 1940s and early 1950s he developed a two-population scheme that ultimately came to be understood in terms of stellar and galactic evolution (fig. 4.1).

The two populations included very different kinds of objects and occupied different parts of galaxies. Population I included very hot, very luminous stars (designated O and B stars), other spectral types of stars, open star clusters, young associations, neutral hydrogen, and other forms of gas and

4.1 This close-up photograph of M31 shows the central bulge of Population II stars and, on either side, Population I, which includes bright stars and dust lanes.

dust clouds.* It was a young population and it seemed to be similar in the abundance of its heavy elements to the sun, being about 99% (by mass) hydrogen and helium and 1% everything else. As originally defined, Population II included only giant red stars plus associated faint stars — no bright blue stars, no gas, no dust — and only globular star clusters. The stars were all old (on the order of 10 billion years old or more) and deficient in heavy elements — typically 10 to 100 times less of the heavier elements (those with masses greater than helium) than the sun. Thus, Population II stars were defined to be almost pure hydrogen and helium.

* There are basically three kinds of star groups found in galaxies. *Stellar associations* are loose, young groupings of recently formed stars, conspicuous because of their very high luminosity and temperature. *Open clusters* are smaller, more compact groups, relatively stable and with a variety of ages. *Globular clusters* are large, very populous, luminous clusters, uniformly very old, generally including the oldest known stars in a galaxy.

Population I was found to exist in the disks of spiral galaxies and in irregular galaxies like the Magellanic Clouds. Population II stars were said to inhabit the central bulges of spiral galaxies and their thin outer haloes. Elliptical galaxies were considered to be pure Population II objects.

Baade's concept of the two populations was simple and astonishingly successful. It led to the unraveling of the complex interplay between age, dynamics, and element production in galaxies, and helped to begin the process of understanding the evolution of stars. But, like many scientific breakthroughs, it eventually was found to be an oversimplification. The grand scheme of two types of stars did not take into account intermediate objects of one sort or another and completely omitted all kinds of exceptions. At first, astronomers were tempted to elaborate on the scheme, defining various subtypes (Population IIa, Population I.5, and so on), but eventually the futility of doing so became apparent. As understanding increased about the properties of galaxies, and as modern equipment improved, the need for a simplified scheme vanished. It served its purpose admirably, but now astronomers can use more quantitative measures of the characteristics of a group of stars. Now an actual age and composition can be assigned to a star cluster or a portion of a galaxy. Baade's Population I and II were used less and less, until today astronomers hardly ever mention them.

One of the reasons the scheme broke down centered on the strong correlation that was originally found between age and chemical composition. Population I was young and rich in heavy elements. Population II was old and poor in heavy elements. This correlation was beautifully explained in 1959, when it was realized that heavy-element production in stars causes a gradual change in the chemical abundances of a stellar population. While the early-forming stars are almost pure hydrogen and helium, those that formed later on, especially those that are forming now, are made up of enriched material, which was produced in the interiors of past generations of stars. Our sun, only 4.6 billion years old, has heavy elements in it that were produced in the cores of previously existing stars. (This is, of course, true for the earth and its inhabitants, as well.) Thus, as the elements were built, stars became richer in these elements and our galaxy could be separated roughly into Population II and Population I stars.

But in the 1960s it was found that the Magellanic Clouds presented a problem. First, these two galaxies seemed to contain many globular clusters that were *young* rather than old. Furthermore, many young stars were found in the

Clouds which, when analyzed in detail spectroscopically, turned out to be poor in heavy elements. Thus, the Population I and II classification did not work for these two galaxies. Moreover, in recent years it has been shown that many of the supposedly pure Population II elliptical galaxies are surprisingly rich in heavy elements, as are the central bulges of spirals. Even in our galaxy there is a breakdown of the scheme; at the very outer parts of the galactic disk the heavy-element abundance in young stars is unexpectedly low, whereas in the globular clusters near the galactic center and in the bulge stars, heavy elements are fairly abundant, despite their advanced age.

Rather than relying on these outmoded categories, we now use specific measures (ages, abundances, dynamics, location) to characterize a population. The following sections describe some of the ways by which we discover these facts.

Star Types

Stars of either population type come with a variety of properties. The principle differences between stars result from their different masses, ages, and chemical compositions. For a typical assemblage of stars, such as that around the sun, the main distinguishing difference, however, is just mass, because most stars are found in a stable, relatively unchanging stage of life and most stars have very nearly the same composition. Mass determines, then, most of the differences we see: high-mass stars are hot, very luminous, and blue in color, while low mass stars are cool, of low luminosity, and red in color. Stars of intermediate mass, such as the sun, are intermediate in their characteristics—fairly cool, of average brightness, and yellow in color.

The age of a star does affect its appearance during certain brief periods of its life history. A star spends most of its life in a state that astronomers refer to as the *main sequence,* during which its color, temperature, luminosity, and other properties remain almost unchanged. But before it reaches this stable state, when it is a *protostar,* it is redder and briefly brighter than its main-sequence condition will be. Unfortunately, protostars are very rarely seen, because that stage takes up only a tiny fraction of a star's life. It is much more likely for us to catch a star when it is in its post–main-sequence stages, which are also short but not as brief as its early life. At the end of the main-sequence stage a typical star becomes a *red giant* (very massive stars become *supergiants*); its color changes, it grows greatly in size, and its luminosity increases. During this period most stars are at their most conspicuous, and we are most apt to be able to detect them in distant environments, such as other galaxies.

When we look at the stars in a nearby galaxy, therefore, we see only the most luminous stars, and these will be the bright blue, massive stars and the red, luminous evolved giants. But the actual life expectancy of a star depends very much on the star's mass; high-mass stars live only a short time, a few million years typically, while low-mass stars can go on shining for billions of years. (The sun has been a main-sequence star, shining steadily, for about 5 billion years and has another 10 billion to go before it turns into a red giant.) For this reason, we will not see bright, blue, massive stars in a galaxy that has not been forming stars lately. For a Population II galaxy, where all the star formation occurred billions of years ago, we will only be able to detect the bright red giants.

Although chemical abundances in most stars are very nearly the same, there are small differences that are of interest. They do not affect the star's luminosity or color very much, but the differences can be detected in the spectra of the stars. Almost all stars are made up mostly of hydrogen and helium; for example, most stars in our area of the galaxy have only about 1% of their mass in the form of other elements. In a few, however, even a smaller percentage of their mass is made up of these heavier elements, down to values as small as 0.01%. We call the stars that are like the sun heavy-element-rich stars, which in our galaxy belong to Population I, while the others are called heavy-element-poor stars, and belong to Population II.

The temperatures of stars greatly affect the appearance of their spectra. Astronomers early in the twentieth century developed a code to distinguish stars with different spectral appearances, which we now know means stars of different temperatures. The hottest stars, with temperatures of 25,000K and greater, are called O stars. Those with temperatures from about 11,000K to 25,000K are B stars, and then come A, F, G, K, and finally M stars, which have temperatures of 3,500K and less. The sun is a G star, with a temperature of about 6,000K at its surface.

Color – Magnitude Diagrams

A handy way to find out what kinds of stars exist in a region of space is to make a plot of their colors and apparent magnitudes. The result is called a color – magnitude diagram. On it one can plot not only the positions of the stars that one observes, but also the expected locations of stars of different masses, ages, and compositions. By comparing the observa-

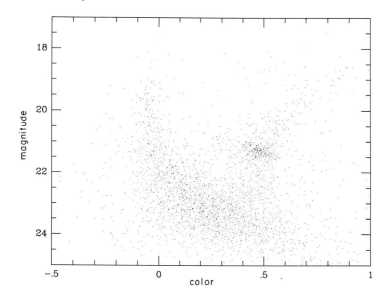

4.2 Color–magnitude diagram for stars in the Large Magellanic Cloud. The vertical axis measures their magnitudes and the horizontal axis their color (from blue stars at the left to red stars at the right). The stars on the left side of this diagram are on the main sequence, while most of the stars on the right are evolved giants.

tions with the theoretical data, astronomers can deduce the properties of the observed collection of stars. For galaxies, only the very brightest stars can be seen, but color-magnitude diagrams of different galaxies nevertheless contain a great deal of information about their stellar populations. In a galaxy like the Large Magellanic Cloud (Chapter 6), for example, we can see the young, blue massive stars on the bright part of the main sequence as well as the red, evolved giant stars, and can figure out the relative numbers of young and old stars (fig. 4.2). As will be demonstrated in following chapters for several individual galaxies, this is a good way for astronomers to deduce things about the past histories of different galaxies (fig. 4.3).

Luminosity Functions for Stars in Galaxies

Galaxies close enough for color-magnitude diagrams to be obtained are also resolvable enough so that the numbers of stars of various brightnesses can be counted. This produces what is called the *stellar luminosity function*. In the neighborhood of the sun, stars are found to be distributed in brightness according to a curve known as the *van Rhijn luminosity function,* which expresses the number of stars within an absolute magnitude interval in a given volume of space. For the solar neighborhood, the van Rhijn function was deduced by a statistical study of nearby stars.

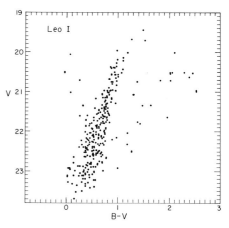

4.3 Color–magnitude diagram for the dwarf elliptical galaxy Leo I. All the stars plotted are red, evolved giants; the main sequence is too faint to detect for this old and distant galaxy.

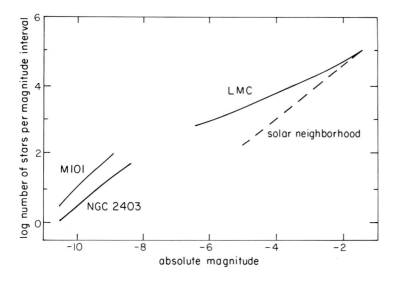

4.4 Luminosity functions for the spiral galaxies NGC 2403 and M101, compared with those of the Large Magellanic Cloud and the solar neighborhood. For the two spirals, which are more distant than the LMC, only the very brightest stars can be counted. For the LMC there are many fewer very bright stars because of its smaller total population. Except for the solar neighborhood curve, all lines represent the total number of stars in the galaxy for each magnitude range.

Luminosity functions are now available for many galaxies in the local group and a few more distant ones (fig. 4.4). For example, the luminosity function for the Small Magellanic Cloud, as measured long ago by Shapley and his associates at Harvard, when compared with the van Rhijn function, shows a relative overabundance of high-luminosity stars. This is probably a general feature of irregular galaxies, which seem to be richer in young, superluminous stars and gas than spiral galaxies like our own.

The luminosity functions of elliptical galaxies, such as the Sculptor galaxy, look much like those for globular star clusters (fig. 4.5). By measuring the surface brightness of the unresolved stars in Sculptor, it has been shown that the faint portion of the luminosity function of this galaxy also must be similar to that of a globular cluster, because the total luminosity of the unresolved portion, as well as its color, is like that of a typical globular star cluster.

Spectral Types of Galaxies

The spectra of galaxies can reveal something about the relative numbers of the different kinds of stars present and can give us some indication of the chemical composition of these stars. The spectra are not simple to interpret, however, because they are composite, and it is not always easy to unravel the effects of the myriads of different stars contributing to the spectrum.

Because of the importance of spectral information to the

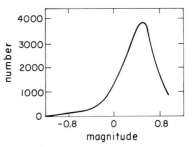

4.5 Luminosity function for an elliptical galaxy, the Sculptor system, as derived by Shapley. The horizontal axis shows the absolute magnitude of the stars in the system and the vertical axis shows the number of stars of that particular absolute magnitude.

problem of the stellar content of galaxies, considerable effort has gone into making spectra highly accurate. When this information is combined with the light and mass distribution in the galaxy and a map of its neutral hydrogen content, a fairly complete model of the content of a galaxy emerges.

Most studies of galaxy spectra use an electronic detector, which accurately measures the brightness in the spectrum at a large number of different wavelengths. The observational data are fitted by computer to various assumed mixes of star types and chemical compositions; and with enough different star types and different chemical line indices, it is possible to derive a good fit which fairly closely defines the properties of the galaxies' stars. For example, for M31 (Chapter 8), the various line indices make it quite clear that the spectrum is composite. Hydrogen lines are conspicuous in the blue part of the spectrum, whereas titanium oxide bands are present in the infrared, implying that there must be some stars hotter than the sun to give the strong hydrogen lines and some stars much cooler than the sun to have the titanium oxide molecules. At the same time, there must be stars roughly similar to the sun because of the strong magnesium line in the green and cyanogen in the blue. The mix of stars in M31, however, is very different from that in the solar neighborhood. There are very many more faint red stars than in our galaxy near the sun. An index of the difference is the ratio of the amount of mass compared with the total light per unit volume, expressed in solar units. The mass-to-light ratio of the solar neighborhood is approximately $M/L = 0.7$, while that for the nucleus of M31 is calculated to be approximately 20.

Gas

Most of the gas between the stars in a galaxy is cool, neutral atomic hydrogen, called HI. Near hot stars the hydrogen is ionized and glows visibly, forming what is referred to as an HII region. The Roman numeral after the H refers to the amount of ionization (the number of electrons lost to the atom because of its hot environment); none is given a I, singly ionized gas is given a II, and so forth. There are small amounts of other gases in space—for example, helium—which have more electrons and can have higher amounts of ionization.

Generally, gas is a very inconspicuous component of elliptical galaxies. Nevertheless, spectrographic surveys show evidence of excited gas in the nuclei of about 15% of them. The intensity of the emission lines ranges smoothly from barely detectable to strong and conspicuous. In dense clusters of galaxies (such as the Perseus and Coma clusters) virtually no emission lines are present, probably because collisions between densely packed galaxies have swept them clean of gas.

As an example of a typical elliptical galaxy for which gaseous emission is detected, consider NGC 4278, an E1 galaxy in the Virgo cluster (fig. 4.6). On spectra taken in such a way that the velocity and the intensity dispersion in the emission lines can be observed at various distances from the center, the emission lines are found to be inclined owing to rotation of the gas cloud. They are extremely broad at the center, because of large turbulent velocities, making the line appear diamond-shaped in the spectrum. Detailed measurements show that the spread of velocity at the center of the galaxy is about 700 kilometers per second. The gas is spread out in a nearly circular distribution around the nucleus. The turbulent velocity decreases outward, becoming nearly zero where the intensity of the emission falls off. The diameter of the gas cloud is approximately 200 parsecs, making it larger than a typical HII region but not larger than the largest that we find in spiral and irregular galaxies. The velocity changes outward from the center at the rate of about 1.5 kilometers per second per parsec. The line at 3727Å of ionized oxygen is 10 times more intense than any others. A comparison of the relative intensities of the different lines shows a resemblance to a typical HII region such as the Orion Nebula in our galaxy. If it is assumed that the temperature of the gas cloud is also similar to that for a diffuse nebula in the Milky Way (approximately 10,000K), then it can be calculated from the line intensities that the hydrogen-to-oxygen ratio is very similar to that of the sun. Because the turbulent motions should be damped out in a relatively short period of time, energy must be fed continuously into the gas in some way. The entire mass of the gas, calculated from its density and its size, is about 100,000 solar masses. We do not know what ionizes the gas. It may be that the high velocity dispersion dissipates enough energy to keep the gas ionized, or it may be that there are a few hot blue stars with enough ultraviolet light to ionize the gas and to give it turbulent motion. The gas cloud most likely is made up of gas ejected by evolving stars, which has fallen to the center of the galaxy.

Surveys of spiral and irregular galaxies indicate that all of them contain gas, both neutral hydrogen and excited gas clouds (fig. 4.7). Sa galaxies show the faintest emission lines, and irregular galaxies have the strongest lines and the highest proportion of gas to stars. The gas fraction in galaxies

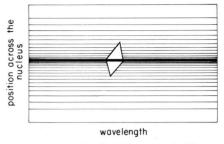

4.6 Diagram showing the remarkable shape of the spectrum of NGC 4278, an E galaxy with a central gas cloud.

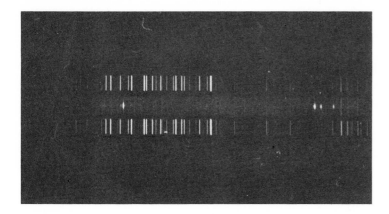

4.7 Photographic spectrum showing strong emission lines in the visual spectrum of the irregular galaxy NGC 5253.

decreases steadily through the progression from Irr to Sc to Sa. Furthermore, the distribution of the gas also depends on the type of galaxy. Gas is often narrowly limited to a ring in Sa and Sb spirals, but widely distributed over the whole galaxy in Sc's and irregulars (figs. 4.8 and 4.9). SBc and SBb galaxies have large amounts of ionized gas in the nuclei, whereas only a few Sc and Sb galaxies do. Generally, the spectra indicate physical conditions similar to those in normal emission nebulae in our galaxy, with temperatures of about 10,000K and electron density less than about 1,000 per cubic cm. The ratio of the strength of the hydrogen lines to those of other elements, such as nitrogen, decreases toward the galaxy's center. Apparently, this is due to a decrease in the ratio of hydrogen to nitrogen, rather than to a change in electron temperature, but other effects may also be involved.

Radio telescopes can be used in three ways to study the gas in galaxies. When tuned to radio wavelengths of 21 cm, they can detect and map the cool neutral hydrogen in the galaxy. Spiral galaxies contain considerable amounts of this gas, spread out relatively smoothly among the stars. It often extends far beyond the outermost visible stars, forming a large, flat envelope of gas centered on the stellar galaxy.

Another way to detect gas with a radio telescope is to look for the radio radiation emitted by hot HII regions. Any hot cloud of gas will emit light of radio wavelengths, but rather than sending out only a particular wavelength (a *line emission*) it emits some light of all wavelengths, and this is called *continuum emission*. The larger and hotter the gas cloud, the more radio continuum radiation it emits. We can map these hot gas clouds in many nearby spiral and irregular galaxies, finding out where the excited gas in them is concentrated.

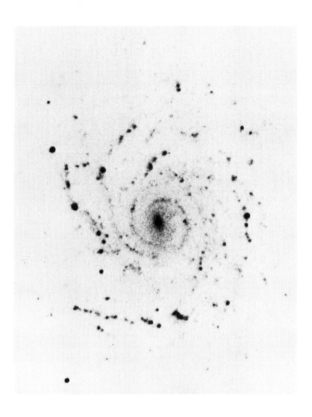

4.8 The distribution of HII regions in NGC 1232, a giant Sc spiral. This photograph was taken with a filter that isolates the light of glowing hydrogen gas.

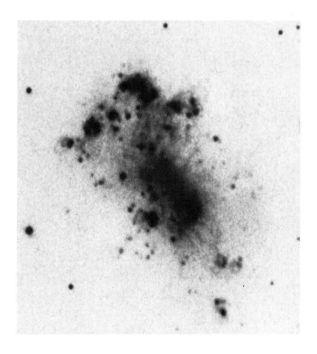

4.9 HII regions in the irregular galaxy NGC 4449, revealed by a hydrogen light photograph.

Finally, a third way to study galactic gas at radio wavelengths is to look at the radiation emitted by high-energy charged particles. It can be distinguished from normal gas-emitted light by its peculiar energy distribution. Called *synchrotron radiation,* this light is only found under special conditions that involve explosive or other violent events, such as found in disturbed galaxies (Chapters 11 and 12).

Distribution of HII Regions

If photographs or electronic images of galaxies are taken with filters that isolate emission lines of hydrogen gas, the distribution of HII regions can be seen. The most conspicuous feature of the HII distribution in spiral galaxies is that these regions occur primarily in the spiral arms. However, they do not exist exclusively there. There are sections, as in the inner portion of one of the spiral arms of M33, where the HII regions are narrowly aligned along the inner edge of the arm, but more generally they are distributed throughout the arm structure, and—especially in galaxies with less than perfectly defined arms—between the arms.

HII regions in spiral galaxies show a maximum density approximately one-third the distance from the center to the edge of the galaxy. There is a tendency for HII regions to be scarce in the center of many spiral galaxies, especially Sa and Sb types, as well as in the outermost arms. In M31, for example, the HII regions are restricted to a ring-shaped area near the position where the arms become dominant over the elliptical component of the galaxy (see also p. 46).

Among the many galaxies for which both HI and HII surveys have been made, there is a great deal of variety. Unlike M31, where the neutral hydrogen and HII regions show peaks at nearly the same position, for most galaxies there is a very conspicuous difference in the position of the maximum neutral hydrogen density and the maximum HII density. The neutral gas usually peaks farther from the center. This is possibly the result of the way in which gas has condensed into stars over the lifetime of the galaxy; probably the most efficient star formation occurs near the center of the galaxy and the completeness of condensation into stars decreases outward from the center.

Irregular galaxies show neutral hydrogen spread throughout their visible disks in a fairly chaotic pattern. In the Magellanic Clouds, for example, it is possible to see neutral hydrogen concentrations within individual stellar associations and even in star clusters. Continuum radio observations have enabled radio astronomers to pinpoint both thermal radio sources, such as HII regions, and nonthermal radio sources, such as supernovae remnants, within many of the nearest galaxies.

Dust

Reliable data on the presence of dust in galaxies is not easy to obtain. The three ways to detect it are through its reddening effect on the colors of a galaxy's individual components (star clusters, variable stars, and O and B stars), through its obscuring effect on the general stellar light of a galaxy and on that of distant galaxies seen behind it, and through the infrared light that the dust itself emits.

In the case of the Magellanic Clouds, color measurements of luminous O and B stars, mostly in associations and clusters, show a general color excess on the order of 0.05 magnitudes, which implies that about 15% of total starlight has been absorbed by the dust. This is a mean value, however, based on measurements carried out in many positions in the Clouds. Much greater reddenings due to much larger concentrations of dust do occur in the Clouds, especially in areas like that around the giant HII region 30 Doradus, where there are also high concentrations of gas.

For some galaxies, such as M31, color measurements of the globular star clusters indicate that light absorption by dust depends on distance from the galactic center. Although photographs show considerable local irregularities in the extinction due to dust in spirals, the distribution usually is symmetrically oriented about the center, decreasing outward and rising in the vicinity of spiral arms.

The method of studying dust in a galaxy by counting distant galaxies seen through it has been applied to several nearby objects, including the very loosely structured dwarf elliptical galaxies in the local group. From the distribution of background galaxies counted over the areas of several of these, it is obvious that there is no detectable absorbing material in them, as the distant galaxies show no decrease in number behind the galaxy.

Distant galaxies behind the Magellanic Clouds can also be detected, and this has provided astronomers with maps of the *total* amount of obscuring matter, especially for the Small Cloud. At the center of that galaxy, according to these studies, there are about 1.2 magnitudes of obscuration. Thus, only about one-third of the light gets all the way through.

Another way to study dust in galaxies is to examine the dust lanes seen in projection against the amorphous back-

4.10 Dust lanes in the edge-on galaxy
NGC 4565, silhouetted against its central
bulge.

4.11 NGC 5194, a giant Sc spiral, and
its small, distorted companion NGC
5195, across which dust from the
extended arm of the spiral can be seen in
silhouette.

ground of the galaxy. For some galaxies, especially those viewed close to edge-on, such lanes are conspicuous components (fig, 4.10). It is clear from these that dust is distributed widely within the spiral pattern of most spiral galaxies and is strongly concentrated to the plane in most cases.

Consider the dust in the spiral arms of the galaxy NGC 5194. This Sc galaxy apparently lies slightly in front of an Irr II galaxy companion, NGC 5195 (fig. 4.11). One of the arms crosses one-half of the image of NGC 5195, and the dust in this arm shows up clearly as an absorption lane crossing the smaller galaxy. Because NGC 5195 is a relatively symmetrical object, it is possible to measure the total amount of extinction caused by dust in the superimposed spiral arm; the galaxy's color directly behind the dust is anomalously faint and red. The excess reddening is very nearly proportional to the total amount of absorption. Furthermore, the dust is distributed over a wider area than the luminous part of the spiral arm. The dust is most conspicuous in the inner parts of the spiral arm, a fact that is also observed in other galaxies. The total absorption caused by the dust, approximately 0.4 magnitudes, is of the same order as the thickness of the dust detected in our own galaxy in the local spiral arm.

The Missing Mass

5

Not many years ago one of the more secure fields in the astronomy of galaxies was the study of their masses. Good methods had been developed to measure them, extensive collections of measurements had been made, and we had lists of values that almost everybody believed. A few worrisome problems arose in the 1960s, particularly regarding the masses derived from looking at the velocities of galaxies in clusters, which seemed to come out too big. But generally it was felt that such easy problems as the mass of the Milky Way or of the Andromeda galaxy were solved.

By 1980, however, the situation took a surprising turn, leaving us at present wholly baffled by the problem of galaxy masses. None of the past answers seem to have been correct, because of a completely unexpected and still not understood complication. Before delving into that mystery, however, we will review the basic methods that astronomers have used to come up with these complex measurements.

Star by Star

It is not difficult to estimate the total mass of a galaxy using very simple assumptions and things that are easy to measure. For example, the mass of our own galaxy can be estimated just from its known radius and the number of stars found near the sun (fig. 5.1). The simple but not very accurate assumptions that we live in an area of typical star density and that our galaxy is roughly a sphere will do the trick. If we

5.1 A telescopic view of the stars in the Milky Way shows that they extend out to many thousands of light years. Dust intervenes, however, and we cannot see all the way to the edge of our galaxy with optical telescopes.

count up the stars in the solar neighborhood and add the mass of gas and dust, we find a density of about 3/100 solar masses per cubic light-year. The radius of a galaxy is about 15 thousand light-years, so the volume, assuming a sphere, is about 13 trillion cubic light-years. The total mass in the sphere is just the volume times the density, so our approximation gives a value of 400 billion suns. This answer is surprisingly close to the values found by more careful means. The density of stars is actually highly variable in our galaxy, and they are certainly not arranged uniformly in a sphere. Nevertheless, the simple act of counting up the stars near us one by one and generalizing the local density gives a good first approximation and an illustrative demonstration of the immensity of our galaxy's mass.

Speeds around the Centers

A much better method of finding a galaxy's mass is based on the rotational motion of the galaxy. The method is not much more complicated than calculating the mass of the sun from the orbital speeds of the planets. If the sun were more massive than it is, our earth would have to go faster around in its orbit or it would be pulled into the sun. A less massive sun with less gravity would mean that the earth should be going more slowly than it does or else it would fly off into deep space. Thus the speed of the earth in its orbit matches exactly what it must be for a stable orbit around a star that has one solar mass.

Similarly, the sun and other stars orbit around the center of our galaxy at speeds that are determined by the galactic mass. If the speed is measured and if the size of the orbit is determined, then the mass that regulates the orbit can be calculated. However, there is one complication. In the solar system almost all the mass is in the sun, at the center of the system, whereas in a galaxy the stars are spread out so that there is still considerable gravitational attraction on most stars from mass that is outside (as opposed to inside) their orbits. This means that the total mass of the galaxy can be determined only by the speeds of stars in the outer parts, where almost all the galaxy is inside the star's orbits. What an astronomer must do is measure the velocities of the stars or other material (commonly the excited gas because it is easiest to measure accurately) all the way from the center to the edge, looking for the point at which the velocities begin to look as though they are responding only to an interior mass. This is called the Keplerian part of the curve, because it was Kepler who found the relation between the planets' speeds and their distances from the sun, a discovery that eventually led Newton to discover the law of gravity. Inside the Keplerian part of the curve, the velocities of stars get larger as we look farther and farther out from the center (fig. 5.2). Finally, they begin to level off and then the velocity curve turns over, the velocities becoming smaller again. Beyond the point of the turnover, the velocities are all Keplerian and they should tell the amount of mass in the galaxy. To be more accurate, astronomers fit the entire set of velocities, measured at all positions, to various mass models of galaxies, thereby learning something about how the mass is distributed as well as what the total is.

This kind of analysis was exploited quite extensively in the 1960s. Astronomers measured the masses of many galaxies and found a relationship between the luminosity of a galaxy and its mass and between the Hubble type and its mass. Generally, galaxies of types Sa and Sb were found to have larger masses per unit of luminosity than those of other types—that is, their stars were dimmer on the average than stars in Sc and Irr galaxies. In all types it looked as if the velocity curve turned down near the limit of the observations. Nature appeared to build galaxies so that we could just

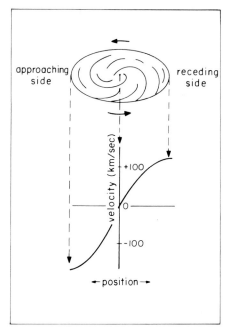

5.2 The rotation curve of a galaxy typically rises from zero in the center to velocities of 100 km/sec or more in the outer parts.

barely see, among their outermost stars, the beginnings of Keplerian motion. The curves fit the mass models well, and the distribution of matter in galaxies looked quite reasonable.

Patterns of Motion

The method of using the circular orbital velocities of stars does not work very well in a galaxy whose stars do not orbit all in one plane. Elliptical galaxies, for example, have stars that orbit around the center in all planes, frequently with highly eccentric orbits (fig. 5.3). There is no flat disk, though there is, at least sometimes, a preferred direction of very slow general rotation. The only ready way to measure the mass of such a galaxy is to examine the *spread* of velocities that the stars show. The speed with which a cloud of stars is buzzing around the center will be sensitive to the total mass, just as the speeds of stars with more circular orbits are. The larger the mass, the larger the average speed of the stars. Some stars will be coming toward us, some going away from us, and some moving across our line of sight, so that a spread of velocities will be seen in the spectrum of such a galaxy (fig. 5.4). The width of this spread, called the *velocity dispersion*, can be used to find the mass of the galaxy, when combined with information on the distribution of stars.

As an example, consider the elliptical galaxy M32, the small, bright companion to the spiral M31. It was one of the first to have its mass determined in this way. The Palomar astronomer Rudolf Minkowski measured the velocity dispersion for its stars in 1952, finding a width of 100 kilometers per second. When the velocity dispersion is combined with the size of M32, one can estimate from this value that the mass must total about 2 billion suns. However, the as-

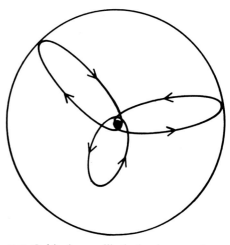

5.3 Orbits in an elliptical galaxy can be highly elongated.

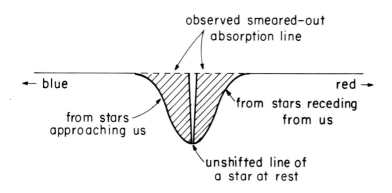

5.4 The velocity dispersion in an elliptical galaxy smears out the spectral lines of the stars because of the Doppler effect.

sumption must be made that all the stars are pretty much the same throughout the galaxy, so that their properties as determined from the spectra near the center will hold true for stars all over. Other assumptions are also made, such as that very slow general rotation is not important, but most of these are relatively minor assumptions compared with the observational difficulties.

Unfortunately, this method has led to some terrible disagreements between different measurements. For M32, for example, results range from Minkowski's original 100 km/sec to several recent measurements of about half that value. An even smaller value comes from using the same method to look at the velocities of nebulae instead of stars. A lot of recent work has focused on the problem of finding more accurate velocity dispersions for elliptical galaxies. About 50 of them have been measured thus far, and they range from about 50 km/sec to almost 400 km/sec.

Once the mass of an elliptical galaxy has been determined with the velocity dispersion technique, the result can be compared with the galaxy's luminosity, and this in turn can tell us something about the kind of matter the galaxy contains. Astronomers express a galaxy's mass-to-light ratio in terms of the mass-to-light ratio of the sun, an average star. Thus, a galaxy made up entirely of suns would have a mass-to-light ratio of exactly 1. Spiral galaxies, it was thought, mostly had mass-to-light ratios of just about that value. Elliptical galaxies, on the other hand, seemed to have values in the range 5 to 10, meaning that these galaxies appeared to contain much matter that is fainter than the sun relative to its mass. Large numbers of dim, lower mass stars, such as red dwarfs, could easily do the trick. While this may indeed be the case in these galaxies, there are other possibilities, including a rather alarming one that is discussed later on in this chapter.

The velocity dispersion technique can also be used to learn something about the stars in the central bulges of spiral galaxies. One cannot learn anything about the total mass of a spiral in this way, but the mass-to-light ratio can be explored. For M31, for example, the central velocity dispersion is found to be about 160 km/sec. The implied mass-to-light ratio is very similar to that for elliptical galaxies, about 10 or so. It seems reasonable to suggest, therefore, that the kinds of stars in M31's bulge are not very different from those that inhabit M32 and other elliptical galaxies. Further evidence tells us that some important differences have resulted from the different chemical and star formation histories of the galaxies, but these are mostly small effects. In general, it would be difficult to distinguish between elliptical galaxies and the central bulges of spirals if we did not see the outer structure of the spirals.

While on the subject of the masses of elliptical galaxies, their interesting dynamics should be given a little more attention. It was pointed out above that the stars' orbits are very eccentric and occupy all the different directions, with no preferred plane (this latter characteristic is called *isotropy*). Until recently astronomers thought that the perfectly round elliptical galaxies were made up of stars with perfect isotropy, no rotation, and a perfectly spherical distribution about the center. Elliptical galaxies that show a more flattened appearance, on the other hand, were thought to be rotating slowly and shaped by their rotation into a flying-saucer shape. To the consternation of the model builders, however, the first actual measures of rotation in the 1970s showed that almost no rotation exists in many "squashed" elliptical galaxies. Some show a slow, gradual rotational motion, but nowhere near enough to explain the galaxies' flattened shape (fig. 5.5).

One possible interpretation of this surprising result is that the galaxies are prolate instead of oblate. A prolate galaxy is shaped like a cigar instead of like a dish, and its shape has nothing to do with rotation in a preferred plane. Another possibility is that the galaxies are oblate all right, but the shape is not simply a result of rotation; it is rather some mix of motions that existed when the galaxy formed and that now show up in the distribution of orbits (somewhat concentrated to a particular plane). If roughly as many stars are going clockwise around the center as are going counterclockwise, the galaxy as a whole will not be rotating, even if the stars are revolving furiously. The total amount of angular momentum could be immense, but the net amount could be very small, as stars rotating in opposite directions would cancel one another out. Which of these possibilities explains the strange slow rotation of the elliptical galaxies is still being disputed. The lesson it teaches us is that objects as huge as galaxies can be much more complex than they look, even when they seem to be pure, simple, and featureless.

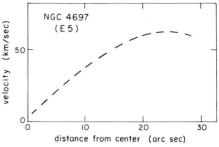

5.5 The rotation curve of an elliptical galaxy shows only small amounts of rotational motion.

Double Galaxies

A still different method of finding masses of galaxies can be used for those that are members of double systems (fig. 5.6). Two galaxies that revolve around each other have to obey

5.6 The double galaxy
NGC 5574-5576.

Newton's law of gravity, which says that their orbits and speeds must depend on their masses. By looking at just one double galaxy, we have no hope of using this fact, because the orbital periods are millions or billions of years, much too long for us to wait. Furthermore, we can only see them from this one direction, so we cannot really determine the angle that their orbit makes with our line of sight. But these problems are overcome if we look at lots of different double galaxies and solve for their properties statistically. Though we cannot follow any given pair around in their orbits, we can look at enough of them to build up an average of their masses.

To correct for the fact that we are looking at two galaxies of very different size when we observe a binary, astronomers calculate the average mass-to-light ratios, rather than the individual masses. This serves to compensate for the fact that the more luminous galaxy will also be the more massive. For example, for a binary galaxy consisting of a very luminous elliptical and a small elliptical, it can be assumed that their mass-to-light ratios are probably the same, but their motions will not be the same. The smaller galaxy will be moving rapidly and the larger slowly around their common center of mass. The average masses deduced will be somewhat in between, not representative of either, but the mass-to-light ratios deduced for the system as a whole will allow the astronomer to calculate the individual masses of each. In actual practice this must be done for many pairs of elliptical galaxies to compensate for the various unknown angles and orbital shapes.

Results of studying pairs of all different types of galaxies are rather surprising. Instead of finding mass-to-light ratios of 1 to 10 — the range for individual galaxies analyzed by the methods mentioned above — astronomers have obtained much larger numbers. Typically, elliptical pairs give values of about 75 and spiral pairs range in the neighborhood of 20 to 40. These values puzzled the people who found them, and were so different from what had been expected that effort was made to find out how they might have gone awry. Were the assumptions wrong in some way? Perhaps galaxies in pairs are inherently heavier (for their brightness) than lone galaxies, for some evolutionary reason. Or perhaps the statistical approach has a flaw in it somewhere. Because of these doubts, astronomers tended to be cautious about the answers that came from double galaxies. They should not have been, but should have saved their worries for the more traditional methods. As the next sections show, the evidence now suggests that the double galaxies were giving us better answers than we realized.

Groups and Clusters

Galaxies in general tend to exist in groups; they congregate. Some, like the Milky Way, belong to small organizations like the local group, while others are members of huge clusters, containing thousands of galaxies (fig. 5.7). In all cases, this fact provides us with another method of finding galactic masses. In a cluster of galaxies, all are moving in accordance with the gravitational pull of the others. How fast they move on the average depends on how far apart they are on the average and how massive they are. The situation is similar to that of the velocity dispersion of stars in a galaxy, but now we are considering the motions of individual galaxies in a cluster. If it is assumed that the clusters of galaxies are stable — that they are not falling into themselves or flying apart — then the motions and separations of their members should give us a measure of their masses.

The problem with this method was that it, too, seemed to give the wrong answer. When average mass-to-light ratios

5.7 The central part of the cluster of galaxies in Pegasus.

were first calculated in this way in the early 1960s, the results were astonishing. Instead of values around 1 to 10, the answers came out in the 100s and even the 1,000s. How could this method be so wrong? Various suggestions put forward included the possibility that the clusters are expanding, that they are contracting, that they consist of very anomalously massive galaxies, that there are lots of double galaxies in clusters (which would lead to higher measured velocities), or that there must be lots of intergalactic matter between the galaxies in clusters, enough to swamp the gravitational field of the galaxies themselves. We now feel a little more cheerful about the cluster results than we did at first. No doubt all of these problems enter in to some extent, but the major explanation is something quite different. All along the galaxies have been hiding a disturbing secret from us: that they are full of mysterious "dark matter."

Dark Matter

Knowledge comes to us in various ways, but the most exciting is known as the breakthrough. This occurs after scientists have been stuck for some time and realize that something is missing; some vital bit of knowledge is there on the threshold, but remains elusive and unfound. The study of galaxy masses went through a phase of this sort when most astronomers felt that there was something wrong with the field, that some important fact had escaped us. The results that were coming in from various ways of measuring masses did not agree well, and the problem was especially acute for clusters of galaxies. The field definitely needed a breakthrough.

The first indication that it was coming was a recent study made of the neutral hydrogen in M31. When gas very distant from the nucleus was detected and measured, the rotation curve refused to turn down and become Keplerian (fig. 5.8). Way beyond where the optical data had suggested that the turndown had been reached, the newer data from the neutral hydrogen indicated that the velocity was actually remaining almost constant. This could only be true if there remained large amounts of mass out in the far reaches of some invisible halo around M31, far beyond the visible parts of the galaxy. All kinds of things were thought of to account for this mass. Perhaps it is very dim red stars, it was suggested, or maybe gas that is ionized so that we cannot see it as neutral hydrogen. But these simple suggestions, like others involving familiar objects, were soon ruled out by various careful observations. The mass out there could not be anything that simple.

In the meantime, other evidence began to turn up which indicated that such massive haloes of invisible matter may be common for galaxies. More sophisticated theoretical models seemed to demand that there be some very massive halo in order to keep the observed flat plane of a spiral galaxy stable. It was argued that the plane would disintegrate unless held in place by the overwhelming gravity of a surrounding mass.

Observers of galaxies other than M31, including our own, began to find that what had appeared to be a turndown in the velocity curve was actually just a small fluctuation in many cases. By the 1980s it began to look as though there were no galaxies whose mass was primarily located within the visible disk. Now we find that a few galaxies do show a Keplerian curve in their outer parts, but the majority do not. Most of the optical and radio rotation curves seem to go on at about a steady velocity right out to the outermost detectable point, even when the most powerful modern techniques are used to record the faintest possible light. Galaxies rarely have most of their matter located within their visible images. Instead,

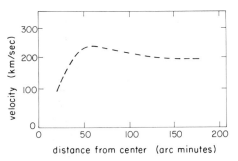

5.8 The neutral hydrogen rotation curve for M31 does not show a turndown in velocity at large distances from the center. Compare with the optical rotation curve, limited to the inner parts of the galaxy, shown in Figure 8.11.

most of a galaxy is beyond the point where we thought it ended.

If galaxies truly have massive haloes, then the strange disagreements discussed above can be understood. The rotation curve method only gives us the amount of mass found inside the outermost measured point (roughly speaking), and the velocity dispersion method only tells us about the mass-to-light ratio in the center, making it necessary to extrapolate outward using the light distribution to find the total mass. Neither method would detect a massive invisible halo. But the double galaxy method would, as the galaxies revolve around each other in orbits that are largely or entirely outside the massive haloes of the individual members. Similarly, the cluster method should be a guage of the total masses of galaxies.

The distressing aspect of this new development is that if these larger measurements are true, then astronomy as it is currently practiced does not detect the majority of the universe. Most of the matter in the cosmos is locked up in some unknown form in the massive galactic haloes, and what we see as galaxies are merely the tips of some very large icebergs. The glorious spiral galaxies are only the skeletons of enormous, mysterious ghosts, the nature of which is still unknown.

Many kinds of things have been proposed to explain the unseen matter in the haloes of galaxies. When physicists first suggested that the tiny particle called the neutrino might actually have a small mass (it had previously been believed to be massless at rest), someone immediately said that the haloes might be made of neutrinos. When it was reported that physicists had detected a monopole (a single isolated magnetic pole) with a tiny mass, someone else immediately suggested that the haloes might be made of monopoles. As other possibilities arose, there always seemed hope that it might explain galactic haloes. Unfortunately, it looks now as if the neutrino may not have mass after all, and the one detected monopole may have been a fluke of some sort, so neither of them will probably solve our problem. We are left with a rather thin list of improbable objects, none of which seems to make a lot of sense. On the list is everything people have been able to think of that has mass but that would be invisible in a galaxy. For example, planets like the earth, if unaccompanied by a shining star, would have mass but would emit too little light to be detected. Smaller objects, boulders or pebbles, for instance, would also qualify. The problem with such things is that no one can think of a way to

produce them in the necessary vast numbers. We are rather
sure that a planet cannot form without a star nearby and the
same goes for boulders. The only reasonable object of this
sort to consider, perhaps, is the black hole, which has mass
and no light, and which might somehow be preferentially
formed in the outer parts of protogalaxies. But whatever it
is — black holes, rocks, or exotic subatomic particles — the
disturbing possibility remains that most of our universe is
hidden from us. We live in a deep, oppressively dark cosmic
cloud, lit only by an occasional candle.

The Magellanic Clouds

6

Although invisible from latitudes much north of the equator, the Clouds of Magellan have been familiar to southern navigators for centuries. Fifteenth-century sailors knew them as the "Cape Clouds" and found them a useful navigational aid. South of the equator, where the north star is perpetually below the horizon and no bright star to mark the south pole can be found, the Magellanic Clouds make a nearly equilateral triangle with the south pole and thus serve as a crude guide. Their use in this way was already common when the great circumnavigator Ferdinand Magellan made his epochal voyage in 1518–1520, during which he perished in the Philippines. When Magellan's ship returned to Europe, his associate and the journey's official recorder, Antonio Pigafetta, suggested that the Cape Clouds be called the Clouds of Magellan as a sort of memorial.

On a dark clear night, away from city lights, a viewer sees the Large Magellanic Cloud as subtending about 5°—that is, about ten times the apparent diameter of the moon—while the Small Cloud appears to be about 2° across (figs. 6.1 and 6.2). This includes only the brightest portions of the Clouds, of course. On sensitive wide-field photographs, one can trace the Large Magellanic Cloud out to a diameter of over 10°, and the Small Cloud to 6° or more.

The total luminosities of the two Clouds, rather difficult to measure because of their large, diffuse extent, are very bright. If the light of the Large Cloud (LMC) were all concentrated in a point, it would appear to be among the dozen

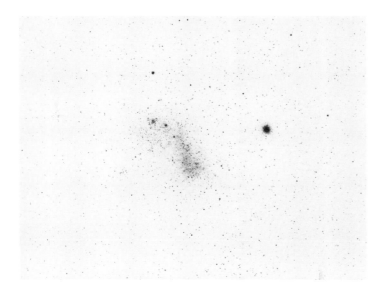

6.1 The Small Magellanic Cloud. This negative photograph shows the faint outer parts of this very close but inconspicuous galaxy.

6.2 The Large Magellanic Cloud.

brightest stars in the sky. The Small Cloud (SMC) is about four times fainter. Both are found to be quite blue in color, owing to the considerable number of very bright, blue stars they contain.

The Magellanic Clouds are fairly typical of their class—irregular galaxies with a minimum of structure. Each has a brightest portion and various irregularly distributed small segments of similarly bright areas. The LMC is dominated by a bright, long, linear structure that resembles the bars seen in barred spiral galaxies. The SMC has a porkchop-shaped central core instead of a bar. Asymmetry is the rule.

Despite their irregularity, the Magellanic Clouds are not chaotic. Both show motions that are relatively well-organized. The LMC, particularly, shows a regular rotational motion that resembles that of spiral galaxies such as the Milky Way. The speed of rotation is slow, reflecting the small overall mass of the galaxy. Masses are estimated to be about 20 billion suns for the LMC and 5 billion for the SMC. By comparison, our galaxy is estimated to have a mass of over 400 billion suns. (These figures refer to the mass of the main bodies of these three galaxies; all may be grossly underestimating the total mass, because of large amounts of unseen matter in the outer parts of galaxies.)

Clusters

The LMC alone is thought to have about 6,500 star clusters, and the SMC nearly 2,000. This wealth of clusters includes a wonderful variety of objects, from giant assemblages of a million stars to faint, tiny aggregates of a dozen faint dwarf stars (fig. 6.3). The Magellanic Clouds, in fact, show an even more profuse range of clusters than does our own Milky Way, a fact that led to early confusion about what kind of cluster was what.

In our galaxy, the globular clusters are usually large, very old, and deficient in heavy elements, while the open clusters are almost always small, relatively young, and like the sun in chemical abundances. Globulars are over 100 light-years across, typically contain almost a million stars, and look like huge, bright swarms of stars, while the open clusters are only 2 to 10 light-years across, contain at most a few hundred stars, and are often too inconspicuous to discover easily.

With the Milky Way's clusters as reference, Harlow Shapley initially divided the clusters in the Magellanic Clouds into these two types; some were clearly globulars but the majority seemed to be tiny, poor open clusters. However, even in the

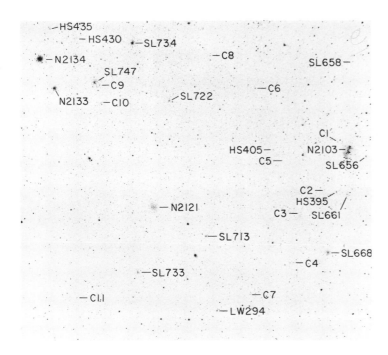

6.3 A portion of the Large Magellanic Cloud in which some of the many star clusters are identified.

1920s there was a confusing observation. Annie Cannon at Harvard had examined the spectra of several of the globular clusters on Shapley's lists and found some of them to be much bluer than globular clusters in the Milky Way. This fact was puzzling at the time because stellar evolution was not yet understood, but we now know that these clusters' spectra indicated the presence of hot, young stars.

Decades later, photoelectric measurements of the Magellanic globulars' color also showed some surprises. The brightest clusters seemed to fall into two groups: "normal" red globulars and abnormal blue globulars. After many more years of research, astronomers now recognize that the blue globulars are young clusters that are otherwise large and populous like the Milky Way's old globular clusters (fig. 6.4). They range in age from infantile objects like the giant NGC 2100, a mere 20 million years old, to clusters of advanced middle age like NGC 1978, about a billion years in age (fig. 6.5). For the older red ones, especially, the difficulties in age-dating the clusters were not overcome until telescopes of 3.6 to 4-meter aperture were put up in the southern hemisphere in the 1970s. Only then was it possible to reach down to the faint, unevolved stars that must be measured to reliably gauge a cluster's age.

The red globulars produced yet another puzzle. As the

6.4 A close-up view of two neighboring clusters in the Large Magellanic Cloud. The one on the right is the young open cluster NGC 1844, while the one on the left, with much fainter stars, is the older cluster NGC 1846.

first color–magnitude diagrams for them were plotted, many were found to be somewhat peculiar in their details, having a less steeply-sloped array of giant stars. When the 1980s saw the introduction of high-gain linear detectors, such as charge-coupled devices (CCDs), which replaced the older photographic techniques, the answer to the puzzle became clear. Many of the supposedly normal globulars are actually somewhat younger than the Milky Way's globulars, though old enough to be red and to look old. Ages of a few billion years are found for such well-studied objects as NGC 2231 in the LMC and NGC 419 in the SMC. Ages for others are still being worked on (it takes months, or sometimes years, to process the immense amount of data that comes from a large modern telescope and a high-resolution, high-gain detector, working on a cluster of thousands of stars). As things stand, only a handful of genuinely old globular clusters have been found in the Clouds; the SMC has one, NGC 121, and the LMC has about six, of which NGC 2257 is a good example. Most of them contain variables of the RR Lyrae type,* as do Milky Way globulars, and have ages measured to be about 15 billion years—about as old as the universe. The rest of the 40 or so large, red clusters appar-

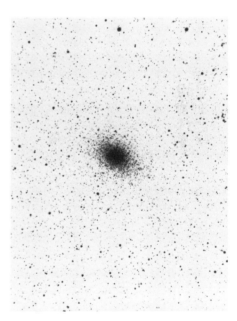

6.5 NGC 1978, an old, highly elliptical cluster in the Large Magellanic Cloud.

---

* RR Lyrae variables are pulsating stars with periods of light variation of a few tenths of a day. They are only found among old, Population II stars.

ently are younger objects. Unlike our galaxy, the Magellanic Clouds have been able to produce giant, rich clusters right up to the present. For reasons not yet thoroughly understood, the Milky Way stopped making them around 15 billion years ago, and since then has just been able to construct small clusters, which in a billion years or less are torn apart by tidal forces and other disruptive effects.

The clusters of the Clouds illustrate the dramatic differences between these two irregular galaxies and our own system. The Magellanic Clouds are loose, low-mass, slowly rotating systems that are not strictly confined to a flat plane and that probably do not have a large spherical halo of old stars, as does the Milky Way. In this more gentle environment, star clusters can form and prosper in a way impossible in the tidally strained environment of our much more massive galaxy.

The richness and range in age of Magellanic clusters make them very useful to theorists. Because Milky Way open clusters are rather poorly populated, having only a few, if any, stars that happen to be passing through the later, giant phases of evolution, we are hampered locally when we want to compare our theories about stellar development with real stars. The rich Magellanic Cloud clusters, on the other hand, provide many stars that are going through their various giant phases, expanding, contracting, and expanding again, before finally collapsing to become invisible as white dwarfs or neutron stars. Thus, the Cloud clusters have shown us what happens in real life, allowing us to compare our theoretical calculations with examples from nature (fig. 6.6). Since such basic facts as the ages of stars are derived from these theoretical predictions, it is of great importance to have a check on the theory. We want to be sure that the various necessary assumptions and approximations that go into any mathematically complex theory such as this have not led us astray. So far, the comparisons with Magellanic Cloud clusters have been remarkably favorable.

The clusters of the Clouds have provided important clues that have helped solve one more puzzle. In our own galaxy, as we have seen, there is a correlation between the ages of clusters and the chemical abundances in their stars. The very oldest clusters, the globulars, tend to have stars that are deficient in elements heavier than helium and hydrogen, such as calcium, iron, and magnesium. Younger stars—for example, stars in open clusters like the Pleiades—have a richer mix of these elements, about the same amounts as in the sun. This is understood as being the result of the gradual

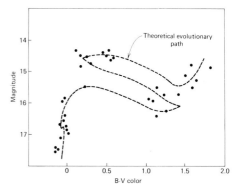

6.6 A comparison of the color–magnitude diagram for the LMC cluster NGC 2164 with a theoretical evolutionary path for a cluster 50 million years old.

enrichment of our galaxy in heavy elements as stars evolve: in its giant phases an evolving star will produce carbon and, perhaps, some nitrogen in its core, while very massive stars that explode into supernovae, destroying themselves in a brief moment of glory as bright as a billion normal suns, produce small amounts of all heavier elements. The longer a galaxy exists and the more supergiants and supernovae it has experienced, the more heavy elements are dispersed into its interstellar medium. Succeeding generations, then, become richer and richer in heavy elements.

For the Clouds, the environment is different and the question arises as to how this might have affected their chemical history. When we look at very young objects, like HII regions or young stars, the spectra suggest that the *present* abundance of heavy elements is less than in the sun. The SMC is low by about a factor five and the LMC by about two or three. Because individual clusters can be age-dated reliably, it is possible to trace how this circumstance has come about. The conclusion is that both Clouds seem to have started out, about 15 billion years ago, with the same low heavy-element abundance that our galaxy had back then. But in the years that followed, heavy elements built up more gradually, especially for the Small Cloud, ending in the presently observed somewhat low values. They appear to be the result of different average rates of star formation and destruction. The lower-mass galaxy, the SMC, processed stars most slowly, while the higher-mass galaxy, the Milky Way, processed them most rapidly. By extrapolation, one might guess that even lower-mass galaxies, such as GR8 or IC 1613, should have even lower present-day abundances of heavy elements, and this does seem to be true. Similarly, very massive galaxies like the brightest seen in the Virgo cluster ought to be unusually rich in heavy elements, a result that also is confirmed by the spectroscopic data available.

The clusters of the Clouds have also provided a means to study its recent evolutionary history. It is not easy to trace the history of a galaxy between its formation and the present, because so much depends upon the wide variety of gravitational and other forces at work in different parts of the galaxy. We must employ methods that are a little bit like archaeology: we start at the present and dig down, layer by layer, to see if a pattern emerges that will give us clues regarding the life history of a galaxy.

The goal is to find the rate of star formation at different times and at different places in a galaxy. One way to do it is to look at the distribution of star clusters of different ages in the

Magellanic Clouds. For the LMC, for example, cluster formation seems to have occurred sporadically and preferentially in groups, with the mean population for groups being approximately 25 clusters, the mean diameter of the groupings of clusters being approximately 5,000 light-years, and the mean duration of enhanced cluster formation being approximately 1 to 2×10^6 years (fig. 6.7).

The Brightest Stars

Much of our own galaxy is hidden from us. Optically, we can penetrate into the thick veil of dust that pervades the Milky Way only a thousand or so light-years in most directions before the absorption and reddening of intervening dust

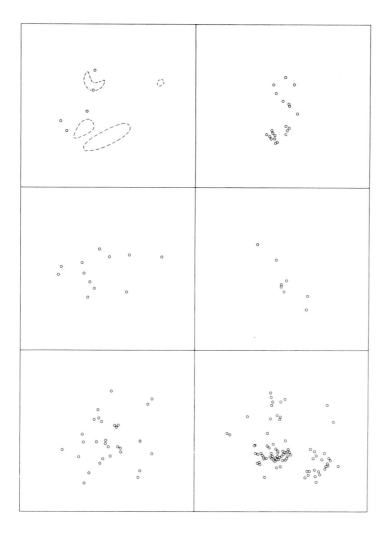

6.7 A sequence of maps showing where clusters formed in the LMC during its recent history. Beginning at the top left, which shows where clusters are forming now, the pictures look back in time at approximately 10-million-year intervals, so that the lower right panel shows the location of clusters formed about 50 million years ago.

clouds obscure more distant stars. For that reason, the Magellanic Clouds are an important source of information
about the rarest types of stars, objects that are too scarce for
there to be many visible in our own system.

The most luminous stars in a galaxy are the rare supergiants, stars that are 100,000 times as luminous as the sun. A
few of these extreme objects have been discerned in the thick
of the Milky Way, but we have nothing like a complete
census. For the Magellanic Clouds, which are spread out
clearly for us to sample, and which have far less dust for us to
contend with, we can see virtually all the supergiants; we can
study them and deduce their histories, their remarkable
makeup, and their life expectancies. We can learn about the
evolution of a galaxy's most massive stars, and we can test the
possibility of using these brilliant objects to gauge the distances to far more remote galaxies, where only the very
brightest stars are resolvable (see Chapter 10).

Modern study of the most luminous supergiants in the
Magellanic Clouds began in the 1950s, and photoelectric
data were combined to discern the membership and properties of this class of objects. It is not a trivial problem. The
stars are so rare that it is not easy to separate them from the
superimposed stars of our own galaxy, which are more numerous in these apparent brightness intervals though intrinsically far less luminous.

How can an astronomer tell whether a given bright star is a
supergiant in the Clouds or merely a normal-sized star in the
foreground, seen by chance in front of the Clouds? The first
clues had come many years ago at Harvard, when Cannon's
spectroscopy in the direction of the Clouds had shown the
presence of very hot young stars, including certain peculiar
stars with envelopes of expanding gas, as indicated by their
spectroscopic profiles. Both of these kinds of stars are among
the most luminous of the local stars and are found only in and
very close to the Milky Way, almost never at positions in the
sky as far from it as the Clouds lie. It could be argued,
therefore, that the stars Cannon had detected must belong
to the Clouds and be among its brightest members. This
method now has been extended to include high-dispersion
spectroscopic study, allowing a direct determination of
whether a given star is a supergiant or not from the width of
its spectral lines. The velocities of hundreds of prospect stars
were measured to weed out those whose small velocities
indicated that they are nearby stars, not Cloud members (the
velocities of the Clouds with respect to the sun are large, 168
km/sec for the SMC and 276 km/sec for the LMC).

6.8 An area of the LMC that contains large numbers of supergiant stars. The cloud-like object is an HII region known as 30 Doradus.

The very brightest stars in the Clouds are actually not the hottest, as might have been expected. Instead, they are A-type stars, with temperatures of about 10,000K. The very hottest stars are 10 times fainter in visual light, emitting most of their radiation at very short wavelengths; these are young, blue stars of types O and B. The coolest supergiants, those with immense, distended outer envelopes of thin gas, are about half as bright as A stars. The most massive stars in a galaxy, those 20 to 100 times as massive as the sun, spend only a fraction of their short lives as hot, main-sequence stars, before they expand into their supergiant, cooler phases.

Recently a great deal of concentrated work has enlarged our knowledge of the Magellanic supergiants. Clever techniques, involving objective prisms at two different positions in front of the telescope, have been used by a group of French astronomers to separate Cloud stars from the foreground by means of their radial velocities. This method can cover hundreds of candidates with only one pair of exposures, and it has increased the census of members to many hundreds for each Cloud (fig. 6.8).

A new tool was added in 1978, when the International Ultraviolet Explorer, a specially equipped orbiting space telescope intended to study the ultraviolet light from hot stars, was first trained on the Cloud's supergiants. Because most of the radiation emitted by hot stars is in the ultraviolet part of the spectrum, it is absorbed by the earth's atmosphere. Thus, to record most of the radiation of very luminous, high-temperature stars, astronomers must use orbiting instruments that view the stars from above the atmosphere. A close ultraviolet examination of the Magellanic Clouds' brightest stars has shown that most are evolved supergiants which were originally two or three times fainter (visually) when they were main-sequence stars. They show evidence of ejected gas from their surfaces, but rather less than in the case of the Milky Way. This indicates that gradual loss of material during the evolution of supergiant stars occurs at a slower rate in the Cloud stars, a probable consequence of their different chemical composition.

Cepheids

Historically, the Magellanic Clouds' most important role has been as a realm for the discovery and testing of the relation between luminosity and period in certain pulsating stars whose brightness varies with time. More than 2,000 variable stars have been found in each Cloud (fig. 6.9), most of them of the type called Cepheids — relatively bright, yellow stars

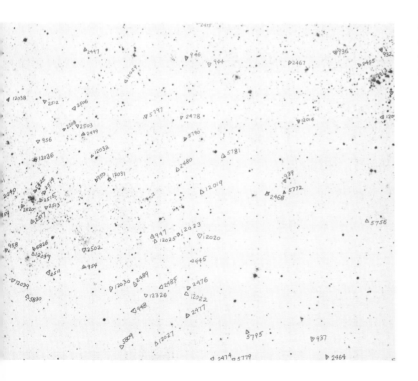

6.9 A section of the Large Magellanic Cloud with the Harvard variables identified by number. Most of the marked stars are Cepheid variables.

with regular variations in brightness on the order of days. Following Henrietta Leavitt's initial discoveries early in the twentieth century, other Harvard astronomers became involved in their study, especially Shapley, who emphasized their importance in distance measurements, and who derived the Cepheid period–luminosity relation for the Clouds. As the data on these stars accumulated, the exact shape of the period luminosity relation was refined (fig. 6.10); recently it has been measured photoelectrically at various wavelengths (fig. 6.11). It is this kind of baseline data on which a distance scale of the universe would eventually be built (Chapter 10). As the nearest galaxies, the Magellanic Clouds are the first vital step in this effort.

Great Clouds of Glowing Gas

On just about any photograph of the Large Magellanic Cloud, one of the most conspicuous features is a bright, sprawling object, located just west of the center of the main bar of stars. This is one of the most massive and luminous hot clouds of gas (HII regions) that we know of, anywhere in the universe. Its popular name is the Tarantula Nebula, and its official name is 30 Doradus (fig. 6.12). Illuminated by a dense cluster of hot stars within it, the Tarantula Nebula glows with the light of thousands of suns. Although similar in nature to the Great Nebula in Orion in the Milky Way, the LMC nebula is immensely larger and more luminous. While the Orion nebula is about 40 light-years across, 30 Doradus extends across a thousand light-years. If it were to lie in our galaxy at the distance of the Orion nebula, it would spread out over the entire constellation and would be bright enough to cast shadows on the Earth. We have no such immense object in the solar neighborhood; only by radio searches have we found any HII regions approaching the size of the Tarantula Nebula, and then only at great distances, hidden deep in the Milky Way.

The total mass of the gas in 30 Doradus is about 5 million times that of the sun. This includes the visible, glowing gas and a surrounding accumulation of neutral gas, detectable only at radio wavelengths. The optical structure of the nebula is marvelously intricate, with interwoven loops and rings extending out from its bright core. Some of these are probably the result of explosions of supernovae and others are likely formed by the intense winds of gas emitted by high-luminosity stars. Embedded in the core of 30 Doradus (fig. 6.13), at the center of its rich cluster of hot supergiants, is a

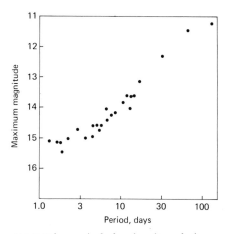

6.10 The period–luminosity relation for Cepheid variables in the Small Magellanic Cloud, as plotted in 1914 by its discoverer, Henrietta Leavitt. The vertical axis plots the maximum apparent magnitude and the horizontal axis plots the period in days.

remarkable object, tremendously bright, called R136a. This star (or stars) is the subject of a current controversy. Astronomers have studied its spectrum using the orbiting International Ultraviolet Explorer telescope, as well as ordinary optical telescopes, and some have concluded that is must be a single, extremely massive star, perhaps 2,000 times the mass of the sun. Others point to optical evidence, including visual studies made by speckle interferometry, that suggests a multiple star — possibly eight stars very close to each other, the largest with masses of 400 or so suns. Until recently, theoreticians who calculated stellar structure claimed that no stable star could exist with a mass of more than about 100 suns, so either possibility qualifies R136a as a most notable object. Once again, nature seems to be saying that our imaginations

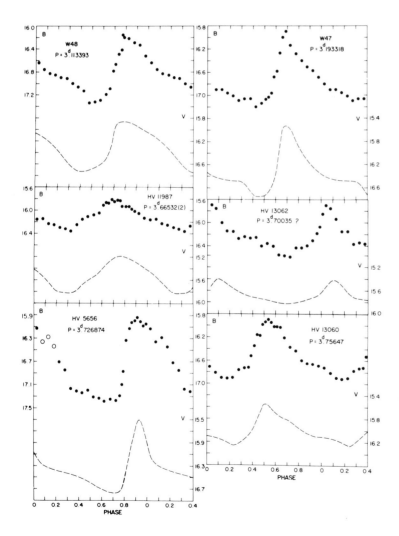

6.11 Light curves of six Large Magellanic Cloud Cepheids, measured in two colors. The lower curve in each case shows the star's variation in brightness in yellow light (labeled V) and the upper curve in blue light (labeled B).

6.12 The brightest and largest gas cloud in the Magellanic Clouds, known as the Tarantula Nebula, or 30 Doradus.

6.13 The center of 30 Doradus, showing the embedded star cluster. The brightest object in the cluster is the remarkable star R136a.

are too limited; the schemes we devise are seldom so grand as what nature has already put together.

There are thousands of other glowing gas clouds in the LMC and hundreds are catalogued for the SMC. Though much less impressive than 30 Doradus, many of these have structures that help astronomers better understand the dynamics of gas clouds in different environments.

The brightest HII region in the SMC is NGC 346, a giant cloud of gas embedded in an unusually rich association of bright stars. It is consistent with the theory that stars usually form in such giant clouds of gas, as the association is very young, just born, and there is still enough gas left over to form a few more stars.

The HII regions of the Clouds provide some information on the chemical abundance of the interstellar medium as well. Compared with gas in the Milky Way (the Orion Nebula, for example), the gas regions in the Clouds are less abundant in heavy elements. Even helium is noticeably deficient, probably the result of there having been fewer generations of stars in the Magellanic Clouds. While in the sun's neighborhood a given atom (such as one inside this page or in your body) might have, on the average, existed at various times in three different stars, in the Magellanic Clouds a typical atom will have had a less interesting past. Perhaps it will have had only one previous stellar incarnation.

The Dilemma of the Dust

Since dust is made largely of heavy elements, it might be supposed that this material should be rarer in the Magellanic Clouds than in our dusty galaxy; and that does indeed seem to be the case. About 50 discrete "dark nebulae," as they are often called, have been catalogued in each Cloud, and they seem to have normal properties when compared with dust clouds in the Milky Way. However, the number of discrete dust clouds is smaller in the MCs than a local sample would lead us to expect, and the overall dust content of both Clouds is clearly anomalously small. As evidence in the SMC, we have the counts of background galaxies (fig. 6.14), whose degree of obscuration indicates the total extent and amount of interstellar dust in the galaxy. To compare various galaxies' dust content, astronomers often refer to their dust-to-gas ratio, as a way of normalizing the data. The SMC's dust-to-gas ratio turns out to be more than three times lower than the Milky Way's.

Confirmation of this conclusion has recently come from a

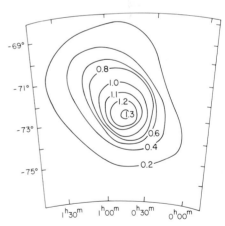

6.14 The distribution of dust in the SMC from background galaxy counts. The figures inside the diagram give the absorption in magnitudes.

series of measurements of carbon monoxide (CO) in the Small Cloud, using radio telescopes. In our galaxy, carbon monoxide is found in giant molecular complexes, which are optically conspicuous for their large dust content (at least, for those that are optically visible; most are buried too deeply in the galactic dust to be detected optically). The CO in the SMC is very, very weak; at the time of this writing, in fact, the question of whether it is detectable at all is still unsettled. Dutch astronomers report CO regions in the bright core of the Cloud, while a team of astronomers at Columbia University and the University of Chile find only very weak CO signals from a few of the optically catalogued dust clouds.

Optical studies of the LMC suggest more dust, but still not anywhere near as much as in our galaxy. Carbon monoxide clouds are also found there, especially in areas where current star formation is prominent and dust is present. The brightest stars and Cepheids of the LMC seem to be reddened only slightly, mostly by dust in the foreground in our galaxy. A few are dimmed by as much as 50% or so, mostly in areas near the center of the Cloud where dust clouds are also conspicuous. However, few galaxies seem to shine through the LMC. There is one bright spiral, NGC 1809, that peeks through the thick curtain of LMC material, but few others can be found. A valuable study for the future will be to determine why the bright stars are so little reddened, while the background galaxies seem to be blotted out almost entirely. Could there be a sheet of dust in the LMC behind all of the bright stars we measure? Or are the distant galaxies difficult to recognize because of a dense bed of faint Magellanic stars in front of them? This remains one of the many unsolved questions raised by the Clouds of Magellan.

The Common Envelope

The hot, glowing clouds of hydrogen are optically conspicuous features of the Magellanic Clouds. But if we could see with long-wavelength eyes, tuned to the 21-centimeter wavelengths that neutral, cool hydrogen gas emits, an even more brilliant view could be had. Radio telescopes have found that both galaxies show HI radiation of immense size and complicated structure, in spite of the rather small total masses of the Clouds. Like other irregular galaxies, an unusually large proportion of their mass is in the form of neutral hydrogen, probably more than 10% in both cases. This seems reasonable, considering that the Magellanic Clouds have been less efficient than our galaxy in forming stars. Therefore, more of the raw materials for star formation are still around.

In the 1950s pioneer Australian radio astronomers found that at 21cm wavelength the Magellanic Clouds are really only one object. An all-encompassing envelope of hydrogen includes both of them in its immense extent. The LMC is a complex, highly structured component on the east side of this envelope, and the SMC makes up a peculiarly double-moded concentration on its west side. The bridge between is thin gas, with few detectable stars.

Australian astronomers have made a more recent and even more surprising discovery. Flung off across the sky is a huge, thin filament of gas, originating at the Clouds and crossing almost to their antipodes. Called the Magellanic Stream, this ribbon of gas seems to connect together several other very low mass galaxies. The most reasonable explanation of the stream is that it is a tidal tail or bridge that was drawn out from the Clouds during a close encounter with the Milky Way long ago. Computer simulations of such an encounter suggest that the Clouds passed into the outer parts of our galaxy in a near cataclysmic event about two billion years ago. The present distances and velocities of the Clouds, as well as the properties of the Magellanic Stream, fit this model well. Whether encounters occurred even longer ago remains a puzzle, as does the question of what is next in store for the Magellanic Clouds, billions of years hence when they may once again come too close.

X-Rays and Black Holes

A final word about the Magellanic Clouds involves their use as important testing grounds for a couple of very modern and startling discoveries. X-ray sources exist throughout our galaxy, but most are faint and their optical sources are not always easy to understand or their distances easy to measure. Several bright x-ray sources have been found in the Magellanic Clouds and these, of course, all have known distances, the distances to the Clouds. Thus, we can determine their optical properties reliably. This is especially important for those that are binary stars, because if we find that the mass of the x-ray star is large, on the order 10 solar masses, then it is probably a black hole, while if it is only two or so times the mass of the sun, it is probably "merely" a neutron star. To know the masses reliably, we must know the distances. The source SMC X-1 was shown many years ago to be a binary with a neutron star, but in 1983 LMC X-4 was shown to be far too massive to be such an "ordinary" exotic object. Instead, scientists concluded that it must contain a massive black hole. The x-rays are emitted by a stream of gas that is inexorably being pulled into the black hole, to disappear forever from our universe. So far LMC X-4 is the best candidate anywhere for a black hole — one more example of how the Clouds of Magellan still serve as signposts in the night.

The Local
Group

7

As the exploration of nearby extragalactic space progressed in the 1920s, galaxies were found to extend on to the limits of all telescopes, but not uniformly in space. Many, perhaps most, galaxies existed in groups or clusters, and ours was clearly no exception. The Milky Way, a giant spiral (similar to the one shown in fig. 7.1), was crowded by two very near neighbors, the Magellanic Clouds, less than 200,000 light-years away. Another giant spiral, M31 (the Andromeda Nebula), was found to lie about 10 times farther, with its various smaller companions and its spiral neighbor, M33. At comparable but slightly lesser distances were the two small, irregular galaxies, NGC 6822 and IC 1613. Altogether, we were surrounded by a grouping of galaxies that seemed loosely clustered and isolated in space, with no other bright galaxies nearby. The nearest examples outside our local family seemed to be fairly remote, several million light-years farther on.

Members

As time provided opportunity, astronomers identified more and more members of our local group. Two distant companions to M31, the faint galaxies NGC 147 and NGC 185, were identified (fig. 7.2). Other dwarf irregular objects were found that might be like NGC 6822 but appeared even less populous and smaller. Seven very faint, tiny galaxies, called dwarf ellipticals, were discovered, one by one, starting in

7.1 The large Sb spiral galaxy M81 in Ursa Major. Our Milky Way galaxy probably looks approximately like this as viewed from a distance. Ours is, however, a little more open and its central bulge is not so pronounced.

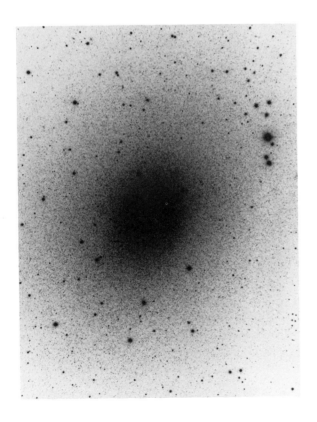

7.2 NGC 185, a companion to M31 whose faint red stars can be seen in this deep photograph. The galaxy lies only a few degrees away from NGC 147, a somewhat less bright companion.

1938 with the discovery of the Sculptor galaxy. The population of our cluster gradually grew until by now we identify
about 20 objects in the local group.

How were these objects recognized for their nearness? It
was not always easy. For giant galaxies like M31, one looked
for Cepheid variables and then applied the period–
luminosity relation to get the distance. But for other kinds of
objects this method did not work. NGC 147, for example,
though we now know it to be just as close as M31, has no
Cepheid variables because its population of stars is wrong; all
of its stellar members are very old stars, among which
Cepheid variables do not exist. Because Cepheids represent
a brief stage in the evolution of massive stars with short
lifetimes, less than 100 million years or so, we can expect to
find Cepheids only in galaxies where star formation has gone
on in the last 100 million years. For NGC 147 there appears
to have been only a single stage of star formation, which
apparently ceased a few billion years ago. Thus, there are
now no Cepheids; some other method must be used to gauge
its distance and to find out whether or not it belongs to the
local group.

For this galaxy as well as its companion, NGC 185, and the
two small galaxies adjacent to M31 (NGC 205 and M32), the
problem was solved by Walter Baade in 1944. Baade recognized that faint galaxies like NGC 147 were Population II
objects, similar in stellar population to globular clusters in
the Milky Way. Shapley had shown long before that the
brightest stars of globular clusters are all very nearly the
same in intrinsic brightness and could be used to obtain
reasonably good distances, and Baade proceeded to use the
same method for Population II galaxies, assuming the
brightest Population II stars all to be similar. This led him to
conclude that all four faint galaxies — NGC 147, NGC 185,
NGC 205, and M32 — must be members of the local group,
all at about the same distance from us as M31.

A different problem arises with certain other candidates
for membership in our cluster. Small irregular galaxies, intrinsically much smaller than the Magellanic Clouds or even
NGC 6822, possibly have too few stars to happen to have any
Cepheids at all, even though they contain other young stars.
Since the Cepheid stage in a star's lifetime is so brief, lasting
only a few tens or hundreds of thousands of years, we could
easily happen to be seeing such a galaxy at a time when no
star is a Cepheid. From a study of its stars, then, we will have
no good criterion for its distance. Are we seeing no Cepheids
because a galaxy is very small and not populous enough to

have any, or because it is larger and farther away, so that its Cepheids are too faint to be discovered? For decades there were a half dozen or so faint galaxies for which this question could not be answered.

Consider a specific example, the strange footprint-shaped galaxy called GR8 (fig. 7.3). It was first catalogued in 1959 at Lick Observatory. Among the many faint images thought to be dwarf galaxies in the Virgo cluster, some 50 million light-years away, No. 8 was a tiny, irregular smudge. A few years later, however, when another astronomer at Lick began a study of the Virgo dwarfs using the 120-inch telescope, he was astounded to see that he could make out visually the brightest stars in GR8. Clearly, this object could not possibly be at the Virgo cluster's distance. For individual stars to be seen visually, without the benefit of long exposures on photographic plates or some other high-sensitivity detector, they must be very close. With a large telescope it is possible to resolve individual stars visually in M31, M33, NGC 6822, and the Magellanic Clouds, but not for somewhat more distant bright galaxies like M81 or the Whirlpool Nebula (M51). Therefore, he reasoned, perhaps GR8 could be close

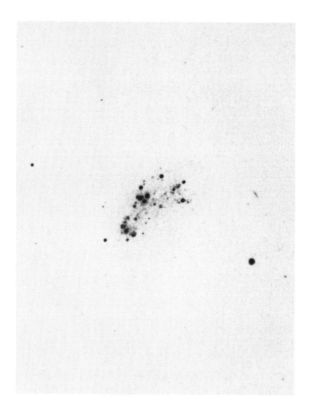

7.3 GR8, an extremely dwarf irregular galaxy in the local group.

enough to be a new member of the local group. Surely it must be at least close to the edge of our cluster, just judging by the crude criterion of its brightest stars. In a subsequent study many plates of GR8 were searched for Cepheids, but none were found. Its tiny size and lack of much nonstellar material suggested that it probably is just too small to have a Cepheid at this random time.

When GR8's velocity with respect to the Milky Way was measured the question of its membership became settled. GR8 and a few other small objects of uncertain distance all behave in motion as would be expected if they were truly associated with our family of galaxies. Other dwarf galaxies, on the other hand, even though fairly close, are moving away from the local group with the general cosmic expansion rate (Chapter 10), and thus do not belong. Galaxies that on this basis are real members are listed in Table 7.1.

Except for the few dwarfs just beyond the local group's boundaries, galaxies are pretty much absent in the surrounding space. The local group is about 3 million light-years across, whereas nearby groups are 10 to 15 million light-years away. Thus, we live in a small community of galaxies, surrounded by thinly populated empty country.

The Irregular Dwarfs

Since it is, after all, our immediate family, it is imortant to spend some time getting acquainted with members of the local group. The Magellanic Clouds, our closest companions, are important enough to have their own separate chapter. For the rest, we will survey them here by type. First, the irregularly shaped, fragment-like dwarfs are discussed; then we turn to the even less conspicuous class of dwarf elliptical galaxies. We save the spirals for another chapter.

NGC 6822
The brightest of the dwarf irregulars is NGC 6822, sometimes known as Barnard's Galaxy. Studied long ago by Hubble, NGC 6822 was the subject of his first paper on nearby galaxies, and in that sense can be considered the first recognized galaxy beyond the Milky Way. Recent work has extended Hubble's research. Newly measured Cepheids provide a new period–luminosity diagram and an improved distance estimate of 1.8 million light-years. Among the galaxy's brighter stars are numerous young, blue stars and a strong population of evolved giant stars.

The very brightest stars are not the hottest, but have tem-

Table 7.1 The local group (in decreasing order of intrinsic brightness).

Name	Type	Right ascension	Declination	Apparent magnitude	Distance (millions of light-years)	Absolute magnitude
M31 (NGC 224)	Sb	00 40	+41	4.4	2.2	−21.6
Milky Way	Sbc	17 42	−28		.03	−20.6
M33 (NGC 598)	Sc	01 31	+30	6.3	2.5	−19.1
Large Magellanic Cloud	Irr	05 24	−69	0.6	0.2	−18.4
Small Magellanic Cloud	Im	00 51	−73	2.8	0.3	−17.0
IC 10	Im	00 17	+59	11.7	4.0	−16.2
NGC 205	E5 pec.	00 37	+41	8.6	2.2	−15.7
M32 (NGC 221)	E2	00 40	+40	9.0	2.2	−15.5
NGC 6822	Irr	19 42	−14	9.3	1.8	−15.1
WLM	Irr	23 59	−15	11.3	2.0	−15.0
IC 5152	Sd	21 59	−51	11.7	2.0	−14.6
NGC 185	E3 pec.	00 36	+48	10.1	2.2	−14.6
IC 1613	Irr	01 02	+01	10.0	2.5	−14.5
NGC 147	E5	00 30	+48	10.4	2.2	−14.4
Leo A	Irr	09 56	+30	12.7	5.0	−13.5
Pegasus	Irr	23 26	+14	12.4	5.0	−13.4
Fornax	E3	02 37	−34	8.5	0.5	−12.9
GR 8	Irr	12 56	+14	14.6	4.0	−11.0
DDO 210	Irr	20 44	−13	15.3	3.0	−11.0
Sagittarius	Irr	19 27	−17	15.6	4.0	−10.6
Sculptor	E3	00 57	−33	9.1	0.3	−10.6
Andromeda I	E3	00 43	+37	14.0	2.2	−10.6
Andromeda III	E5	00 32	+36	14.0	2.2	−10.6
Andromeda II	E2	01 13	+33	14.0	2.2	−10.6
Pisces (LGS 3)	Irr	01 01	+21	15.5	3.0	− 9.7
Leo I	E3	10 05	+12	11.8	0.6	− 9.6
Leo II	E0	11 10	+22	12.3	0.6	− 9.2
Ursa Minor	E5	15 08	+67	11.6	0.3	− 8.2
Draco	E3	17 19	+57	12.0	0.3	− 8.0
Carina	E4	06 40	−50	(>13.0)	0.3	− 5.5

peratures of about 20,000K. They are not as luminous as the Magellanic Clouds' brightest stars, a probable result of the galaxy's overall smaller population. The red supergiants, however, are about the same luminosity as those in the Magellanic Clouds. This is because stars of a variety of masses tend to evolve into red supergiants of about the same maximum luminosity.

NGC 6822 has 16 bright stellar associations, averaging about 500 light-years in diameter (figs. 7.4 and 7.5). Star clusters are also present, with 30 clusters identified so far,

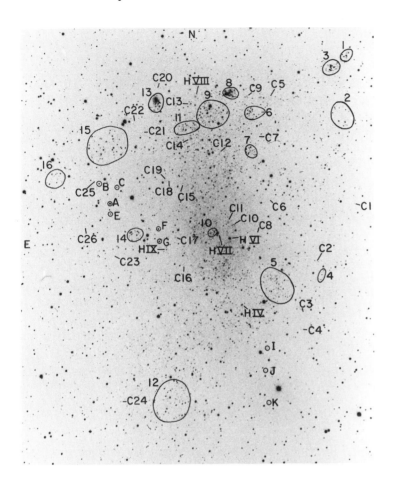

7.4 The stellar associations and star clusters of NGC 6822. Numbered and outlined objects are young stellar associations; objects with Roman numerals and with numbers prefixed by a C are star clusters. A few standard-brightness stars are lettered.

ranging in age from very young to at least a billion years. No globular clusters have been identified for certain, though one or two of the five clusters Hubble found may be old enough to be so classified. They are small, though much larger and brighter than the other clusters. With its small total population, NGC 6822 should not be expected to have any giant globular clusters: the SMC, after all, has only one or two true globulars.

Among the youngest objects in a galaxy, the HII regions are often the most obvious. This is certainly true of NGC 6822, whose brightest gas clouds were the only things that some of its first observers could find. Hubble found five, and subsequent searches have turned up many more; about 40 are now known, and there are probably several more. The biggest and brightest are beautifully structured complexes of gas and hot stars, smaller than 30 Doradus but not unlike it in structure. They are bright enough to allow detailed

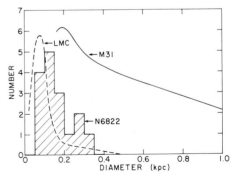

7.5 The sizes of the stellar associations of NGC 6822, compared with those in other galaxies. The vertical axis tells how many associations are found in each size interval.

spectroscopic study. Using this method, astronomers at the University of Mexico, for instance, have found evidence that both helium and all detectable heavier elements are lower in abundance than in our galaxy and the Magellanic Clouds.

In 1980 two Rice University astronomers discovered several possible planetary nebulae* in NGC 6822. Using the Cerro Tololo Interamerican Observatory in Chile to analyze the spectrum of one of them, they found high temperatures (about 18,000°K) and anomalously large nitrogen abundances, probably due to the production of this element in the core of the star before it collapsed and formed the nebula surrounding it. By contrast, the galaxy's HII regions show a comparatively low nitrogen abundance. The ratio of nitrogen to oxygen seems to tell how much nuclear processing has gone on in a galaxy. If the ratio is low, there has been less stellar evolution. NGC 6822 has one of the lowest nitrogen-to-oxygen ratios found anywhere. In time, however, evolution brings up the nitrogen content, as the planetary nebula shows. The N-rich gas is expelled from the dying star and gradually puffs away, to mix with the interstellar gas. New HII regions and stars forming from them will consequently have a higher nitrogen content.

Neutral gas in NGC 6822 has been studied by radio telescopes, though in less detail than the optically shining gas, mainly because of interference from the foreground hydrogen in our own galaxy. A huge envelope of thin, slowly revolving gas was found to extend far beyond the optically visible part of NGC 6822. Inside it is a bright, patchy arrangement of gas coinciding with the galaxy's optical core. High concentrations of gas occur primarily where bright OB stellar associations are found. For example, the nebula Hubble X, which is at the core of stellar association A13, coincides with a giant cloud of neutral gas containing about 2 million times the sun's mass. The rotation of the galaxy's core shows up clearly on the HI maps; the velocities vary fairly smoothly from the northwest corner to the southeast corner, with a range of about 75 km/sec. The total mass derived from this study is 1.4 billion suns; this refers to all material inside a radius of 8.5 thousand light-years.

The optical structure of NGC 6822 is fairly simple, and was described qualitatively by Hubble. There is a vertical bar-like core with a crossing of young objects (HII regions,

---

* Planetary nebulae are small gas spheres surrounding highly evolved stars. They consist of the ejected outer envelopes of gas of the dying star.

stellar associations, and HI gas) above it, shaped like the letter T. Less conspicuous is a broad tail of material that swings off to the east from the bottom of the bar. In gross structure, NGC 6822 is similar to the LMC; patches of furiously active star-forming regions are scattered across a basic bar-like assemblage of older stars, with the most prominent concentrated near one side of the bar (fig. 7.6). Attempts to dig back into the history of star formation in both galaxies show that the same pattern is prevalent as far back as can be traced. Age-dating the star clusters and stellar associations, we find that star-forming regions have bounced around from one place to another in a nearly random way in the galaxy's recent history. The rate of cluster formation is quite low — on the average, one every 6 million years. The LMC, by comparison, manages to form a cluster every 30,000 years.

IC 1613

We have devoted several pages to NGC 6822, not because it is a particularly noteworthy member of the local group but because it is a good example of a dwarf galaxy that has been well studied. Similar work has been done on other dwarf irregular galaxies in our neighborhood, most notably on IC 1613 (fig. 7.7), a faint galaxy at about the same distance from us as NGC 6822. While NGC 6822 is structurally somewhat like the LMC, only smaller, it could be said that IC 1613 resembles the SMC, but also smaller. Its brightest stars are fainter, its HII regions less massive, its star clusters almost invisible, and its Cepheids few in number; otherwise all are normal. The only abundant commodity in IC 1613 is neutral gas, which makes up almost 20% of its total mass, a value about as large as found for any normal galaxy yet studied. Hydrogen gas amounting to 70 million times the mass of the sun is spread over IC 1613's pale face.

A long-standing puzzle has to do with star clusters in IC 1613. Where are they? Walter Baade first pointed out in the 1950s that in all the many years he had studied this galaxy, he had found no clusters at all. The galaxy's mass is about one-third that of NGC 6822 (about 400 million suns), and yet there is no sign of any bright clusters like those discovered by Hubble in NGC 6822. Clearly, we are being told that some agent, yet unrecognized, governs whether or not a galaxy can form rich clusters.

In 1978 it was found that there are, indeed, a few very faint objects that seem to be puny little star clusters, previously missed because of their faintness and tiny size (fig.

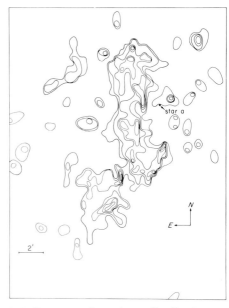

7.6 Contours of equal star distribution in NGC 6822.

7.7 IC 1613, a faint irregular galaxy in the local group.

7.8). Of 43 candidate objects, only 25 seem at all certain. Several are loose aggregates of half a dozen resolved bright blue stars—young clusters of doubtful future—and four are old clusters, whose brightest red giants are just barely resolved on the very best Palomar 200-inch plates. So the clusters of this anemic star system have been found, but we remain puzzled as to why they are so tiny, so faint, and so minor a component of the galaxy.

Since IC 1613 is a very small galaxy, one is not surprised to find that it, like NGC 6822, is deficient in heavy elements as compared with the Milky Way. Present data indicate that IC 1613 is at least as low in these elements as NGC 6822, with a concomitant dearth of dust. Only 11 tiny dust clouds have been discovered, the smallest number in any galaxy searched for these dark, obscuring objects. Background galaxy counts indicate only a very minor obscuring effect. At most, only 50% of faint background galaxies near the center of IC 1613 are screened by a general layer of dust. For the SMC, a comparable search shows an 85% decrease; therefore, there is relatively less dust in IC 1613 than even in the dust-poor SMC.

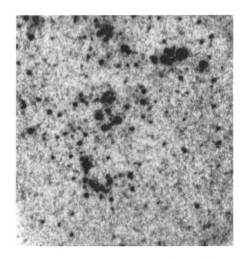

7.8 Three faint, young star clusters in IC 1613.

Other Irregulars

GR8, an extremely small member of the local group, has already been mentioned. Its remarkably small size (about 1,500 light-years) and its low absolute luminosity (only a little more than the luminosity of the brightest single star in a normal galaxy) attest to its dwarfishness. A similar mini-galaxy is LGS-3, a low-surface-brightness galaxy found in 1978. Its distance is unknown, probably somewhere in the range of 0.5 to 2 million light-years.

There are also several dwarf irregular galaxies near the local group, but not clearly close enough to be members. These are all faint objects, discovered too recently to have NGC numbers and mostly known by their home constellation, such as Phoenix, Leo A, and Leo B. The system called WLM was named after its discoverers, Wolf, Lundmark, and Melotte, while IC 10 was found long enough ago to make it into the *Index Catalogue,* a supplement to the *NGC.*

Companions to Andromeda

The Andromeda Galaxy (M31), like the Milky Way, is surrounded by a family of dwarf galaxies. Rather than two close irregular galaxies, like the Magellanic Clouds, the nearest companions of M31 are two elliptical galaxies, M32 and NGC 205. From our vantage point, both are superimposed onto the image of M31, with M32 appearing buried in the south central outer spiral arms and the larger, fainter NGC 205 set among the faint, outer stars of the main galaxy's northern regions. Two similar elliptical galaxies, only a little fainter than these, are found about six degrees away. They are NGC 147 and NGC 185, close enough to each other to make a probable binary pair. The rest of Andromeda's known companions, found in 1970, are four very faint dwarf galaxies, usually called And Dwfs I through IV. Three of them are dwarf ellipticals and one is a faint irregular galaxy.

M32 is a moderately bright, quite compact galaxy, with a total luminosity about that of the SMC, but with its light all packed into a nearly circular, smooth, bright image 6,000 light-years across (fig. 7.9). It is primarily made up of intrinsically faint stars, mostly old stars somewhat like those of globular clusters. However, its spectrum and color indicate that its stars have chemical abundances different from those of the old, heavy-element-poor globular clusters, as was expected from its red color. Instead, there is a population of metal-rich stars, together with a component of only moderately old stars (probably only 2 or 3 billion years old) making

7.9 M32, a small but bright elliptical companion to M31.

up a minor contaminant of the older population. Sparsely scattered among these stars are several planetary nebulae, but no other nonstellar objects have been found. There are no great glowing gas clouds, no dust lanes, no neutral hydrogen gas—none of the things that a galaxy must have to form new stars. M32 is an elderly object with some middle aged parts, but no signs of youth.

M32 is one of the most thoroughly observed galaxies, spectroscopically. Its high surface brightness has made it a favorite target for optical astronomers, who can get data for it rapidly; thus it has become a ready, easy, comparison object for many galaxy surveys. The mass estimate of M32 is derived from its range of observed velocities. Its spectral lines are widened by the Doppler shift, which shows that the full range of velocities near the center of M32 is about 140 km/sec. Thus, some stars are moving away from the nucleus at 70 km/sec and some are coming toward us at that velocity. From these velocities and the stars' distances from its center, we can calculate that the main body of M32 has a mass of about 2 billion (2×10^9) solar masses, making it less than 1% of the mass of its "parent" galaxy, M31.

Even though its mass is small, M32 is apparently responsible for a considerable amount of disturbance. M31's fragile spiral structure shows a confusing and complex, irregular pattern near M32. The arms of neutral hydrogen gas are displaced from the arms of stars by some 4,000 light-years and it is impossible to trace the arms continuously through this disturbed area. Computer models of M31 show that a distorted pattern much like that seen can be produced by a close passage of a companion galaxy with the mass of M32.

Probably, however, Andromeda has had its revenge on M32. Multicolor data for this compact galaxy seem to indicate that its stellar population is like that of a much more luminous elliptical galaxy. Because its compactness is also unusual for a galaxy of its luminosity, possibly M32 was once a much larger object, before close encounters with M31 pulled away its outer stars. Maybe these eventually mingled with the stars of M31, leaving only a dense core behind.

The Andromeda Galaxy's other close companion, NGC 205, has some peculiarities of its own. On photographs it looks like an ordinary elliptical galaxy, rather elongated, of type E5. It is not particularly luminous, only a little brighter than M32. Though twice the diameter of M32, NGC 205 is still small by comparison with M31, which is more than 10 times larger yet. But both the shape and content of NGC 205 are known to be somewhat strange. Elliptical galaxies are normally symmetrical, with isophotes that are concentric, similar ellipses. For NGC 205, however, the very outer regions show a marked twist, with the major axis of the image making a fairly sudden change of direction. This kind of peculiarity is not unheard of for elliptical galaxies; in fact, it is thought in other cases to be an optical effect that results from our viewing a somewhat prolate galaxy from a particular angle. But for NGC 205, the twist is different; it only shows up in the very outer isophotes, rather than being a gradual shifting of the axis. In that way it more nearly resembles the tilting of the outer plane of the disks of our galaxy and M31 (Chapter 8), which is probably (though not certainly) a tidal effect on these thinly populated areas caused by the companion galaxies. The outer plane of NGC 205, though not a flat disk like that of a spiral galaxy, could also be responding to tidal forces. Thus, there seems to be evidence that the galaxies in the local group are interacting with each other, causing distortions, tearing pieces of each other away, and producing peculiarities of structure that can help us to reconstruct their past.

The content of NGC 205 is its other anomalous property. Whereas elliptical galaxies are supposed to consist entirely of

7.10 Dust lanes in NGC 205.

old stars, with no interstellar material, NGC 205 was found by Walter Baade in 1944 to have a small population of very young stars, as well as some conspicuous dust lanes (fig. 7.10). Recent deep probes have turned up hot O and B stars, luminous carbon stars, and possible Cepheid variables (fig. 7.11). Radio astronomers have also discovered a large thin cloud of neutral hydrogen gas in it. Thus, NGC 205 has the raw materials to form new stars, as well as the inclination to do so.

Why is this particular elliptical galaxy going against the rules? Why should it have gas and dust, and newly formed stars, when every other property suggests a "dead" elliptical galaxy, of the sort that supposedly exhausted all of its stellar building blocks eons ago? One possible explanation lies in its proximity to M31; somehow, perhaps because of tidal inter-actions, a bit of M31's vast supply of gas and dust got trans-ferred to NGC 205, where it fell down into the center and proceeded to start forming a new population of stars. This hypothesis sounds plausible until one considers the strange case of NGC 185.

Like so many galaxies, NGC 185 was first catalogued in

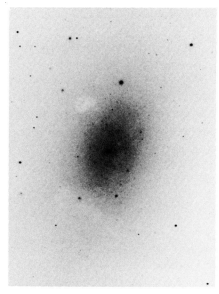

7.11 The bright blue stars in NGC 205 are the conspicuous objects near the center of the galaxy in this blue photograph.

the eighteenth century by Sir William Herschel. In the nine-teenth century a nearby, fainter, but rather similar galaxy, NGC 147, was discovered by H. D'Arrest. By the twentieth century it was realized that these two objects are elliptical galaxies of rather faint absolute luminosity and that they form a pair of distant companions to M31, some 400,000 light-years distant (about twice as far as the Magellanic Clouds are from the Milky Way).

NGC 185 is most notable for its contamination (fig. 7.12). Though somewhat smaller and fainter than NGC 205, it otherwise is rather similar. Instead of a pure population of old stars, one finds a small proportion of young stars, gas, and dust. The two dust clouds make up a patchy, nearly circular shell centered on the galaxy, about 200 light-years from its center. The young, blue stars are spread over the face of the galaxy, also near the center, but the neutral hydrogen is concentrated to one side. A possible interpreta-tion of this remarkable situation is that NGC 185 held back some of its gas at the time of most of its star formation, some 10 billion years ago. That gas is only now being used to form stars. The stellar winds from the concentration of stars near the center of the galaxy, though low, may be enough to explain the nearly circular shell of dust; the starlight pushes out against the dust, preventing it from falling to the gravi-tational center. The residual gas and the dust is small in amount (only a few million suns in mass), but enough to form stars at a slow rate, or, perhaps, in only a recent small burst. Only about 100 young stars are detectable on deep photo-graphs of NGC 185.

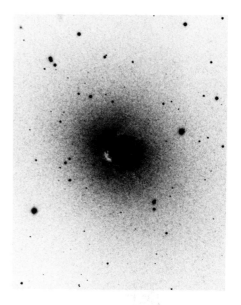

7.12 NGC 185, with its conspicuous dust lanes and bright blue stars.

The total mass of young stars, gas, and dust in NGC 185 is remarkably similar to that found in NGC 205, so that any explanation of this material should be equally applicable to both galaxies. A hypothesis that depends on NGC 205's proximity to M31 will not work for NGC 185, which is much too far from M31, and thus is not very attractive. It is now argued that the contamination by young stars is most proba-bly a perfectly normal feature of elliptical galaxies. These stars are thought to be formed from leftover gas or gas ejected by stars through normal stellar evolution, and are probably found in *all* elliptical galaxies. The spectra of ellip-tical galaxies do suggest the faint presence of a young, blue component, the relative importance of which increases with decreasing luminosity. Therefore, it seems plausible that there may always be a modest amount of star-forming activ-ity, even in supposedly ancient galaxies like these ellipticals. Perhaps the formation rate is discontinuous; bursts of star formation are characteristic of other kinds of galaxies. Maybe elliptical galaxies also have bursts, but the number of stars formed during any one burst is very small. If so, the number does not scale with the mass of the galaxy; that is, very massive ellipticals do not have proportionally massive bursts of star formation. Were that the case, the colors of low-mass galaxies would not be "abnormally" blue, as is observed. This blueness is absent for the massive galaxies because the new stars' light is swamped by the immense numbers of old stars.

NGC 147 is even fainter than NGC 185 and is even more inconspicuous (fig. 7.13). Like NGC 185, it has a few globular clusters, but no gas, dust, or young stars have been detected. Though there is no evidence, therefore, for very recent star formation, the presence of carbon stars indicates that it is not exclusively made up of ancient stars. At least some star formation must have happened within the last billion years or so.

The galaxies known as And I, II, and III are remarkable objects, very much like the nearer Sculptor-type galaxies in almost every way. Astronomers recently have measured luminosities and colors of their brightest stars, a marvelous feat, considering their faintness. This kind of research only became possible in the 1980s, when extremely sensitive, linear, digital detectors, such as the CCD, became available for use on large telescopes. The Kitt Peak National Observatory's 4-meter telescope with a CCD detector was used to obtain data on the And dwarfs, showing that they contain old, heavy-element-poor giant stars.

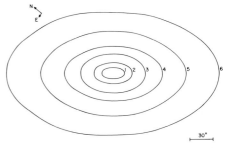

7.13 NGC 147 isophotes, illustrating the galaxy's near perfect elliptical shape.

The Seven Dwarfs

Harlow Shapley reported the discovery of the first of the nearby dwarf elliptical galaxies of the local group in 1938. He did not know what it was, but recognized it as something new, different, and important. One of Shapley's assistants had been searching for galaxies on a series of deep photographic plates taken with Harvard's 24-inch survey telescope at its Boyden Observatory in South Africa. Among the thousands of tiny galaxy images that were being identified was a peculiar image, at first thought to be a fingerprint or other kind of blemish. When it was brought to Shapley's attention, he looked at other plates of the region, which, despite their inferiority, showed just the slightest traces of this strange object. It was real. It occupied much of the upper part of the plate and seemed to consist of many thousands of faint, almost invisible stars, all just above the plate's limit of resolution. Careful counts of the images showed that they were spread out nearly uniformly, filling up a circular area with a diameter nearly four times that of the full moon.

When Shapley published his description of this strange object, he still did not know what to make of it. He called it "the System in Sculptor," after the southern constellation in which it was found, but was unable to decide whether it was an unusual, faint cluster of stars in our Galaxy, a separate galaxy of faint stars, or perhaps a far distant cluster of galaxies. In any case, it was unlike any previously found example, whichever it was (fig. 7.14).

He followed up the discovery with three projects designed to answer the puzzle of the nature of "Sculptor," as it eventually was called for short. First, he assigned an assistant the task of examining *all* Harvard plates of the area, which went back to the 1890s, when the University first began surveying the southern skies from Peru. He wanted to know whether Sculptor could be found on plates taken with other, smaller telescopes. Second, he set assistants the much more time-consuming task of searching all of the 24-inch plates, Harvard's deepest survey plates, to find any additional objects of this sort. And third, he contacted the staff of the Boyden Observatory, requesting that plates of the Sculptor system be obtained with the recently installed 60-inch reflecting telescope, in the hope that its larger scale and fainter limits

7.14 The Sculptor system.

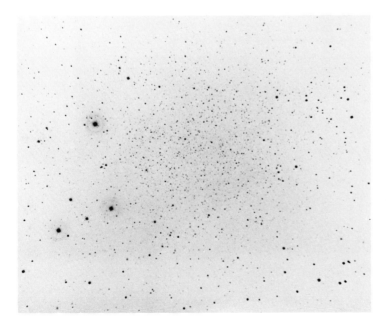

might help determine the true nature of this enigmatic object.

All three projects bore fruit. A plate taken many years before with a 3-inch telescope, exposed over several nights, was found on which Sculptor showed up as a faint, unresolved smudge. By eye, Shapley estimated that its total luminosity must be about that of a 9th magnitude star, but it was spread out over such a large area that its surface brightness was extremely low.

Eventually, a second Sculptor-type object was found on the Harvard 24-inch telescope plates, this one made up of unresolved stars spread out over about the same amount of sky, but in the constellation Fornax. On the plates it looked something like the image of Sculptor on smaller telescopes' plates, and Shapley suspected that is must be the same kind of object, but farther away (fig. 7.15).

When the 60-inch telescope's plates were taken in South Africa and shipped back to Harvard, Shapley and his colleagues eagerly examined them. Clearly, the increased power of the new telescope showed that the images must be

7.15 The Fornax galaxy, displaying its brightest stars and one of its globular clusters (the bright object near the center of the picture). The bright star at bottom center is a foreground star in our galaxy.

stars. More exciting, a few variable stars were found and studied by the Harvard variable star expert, Henrietta Swope, who determined periods and light curves. These objects were identified as Cepheids and immediately told Shapley the answer to his questions; when he applied the period–luminosity relation to them, he found that Sculptor must be an external galaxy out among the members of our local group.

In the meantime, Edwin Hubble and Walter Baade had turned the Mt. Wilson 100-inch telescope almost on its side in order to get a look at these southern objects (both are at declinations of about $-35°$). A series of plates, none very good because of the terrible seeing that always occurs near the horizon, revealed about 40 RR Lyrae variables in Sculptor, and its distance based on these objects was calculated to be about 200,000 light-years. The Fornax system was resolved into fainter stars, none brighter than about magnitude 21, and therefore, if similar to Sculptor, it must be at a distance about twice as large.

The Mt. Wilson astronomers, furthermore, noted the similarity of the Sculptor and Fornax objects to ordinary globular clusters and suggested that they were not really anything new or special, just very big, low-density, globular-cluster-like galaxies. It is historically interesting that Baade later expressed regret at having taken that point of view. If he and Hubble had thought about these galaxies differently, they might have hit upon the connection between globular clusters and elliptical galaxies that later led Baade to formulate his famous Population I and Population II dichotomy, which paved the way for our present understanding of stellar evolution.

For the next 40 years the Sculptor and Fornax galaxies remained examples of what was thought to be a pure globular cluster population of stars, in the form of very spread out, low-density galaxies, which have come to be called "dwarf elliptical" or "dwarf spheroidal" galaxies. Five more examples have been found in the local group, four of them the result of searches of plates from the Palomar 48-inch telescope in the 1950s, when it was first making its deep survey of the northern skies (figs. 7.16 and 7.17). The seventh (Carina) was found with the UK Schmidt, the 48-inch survey telescope in Australia, in 1977.

Something of Baade and Hubble's initial lack of enthusiasm characterized the study of these objects for much of their recorded history. They were regarded as "merely" overblown globular clusters by most students of galaxies. In

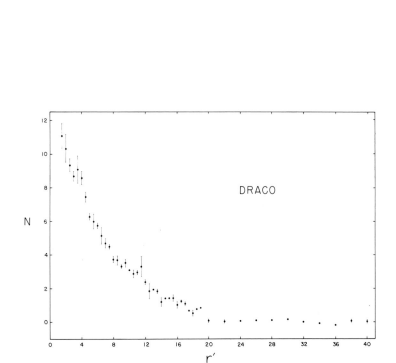

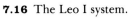

7.16 The Leo I system.

7.17 The distribution of stars in Draco. The density of stars is given by the vertical axis and the distance from the center of the galaxy is given by the horizontal axis. The bars shown for each data point indicate the statistical error of the point.

the 1980s, however, a surge of interest has developed, primarily because it has been found that there are several dissimilarities between the seven dwarfs and normal globular clusters. The first to be noticed is the presence of unusual variable stars, now referred to usually as "anomalous Cepheids." The variables that Swope studied in Sculptor are examples. They are like RR Lyrae variables in some ways, but have longer periods (around 1 to 3 days) and are brighter. One of the dwarfs, Leo I, has at least 20 of these anomalous Cepheids. Most astronomers believe that they are an indication that the dwarf elliptical galaxies include a *younger* population than the globular cluster stars, which are all about 15 billion years old. The anomalous Cepheids have calculated masses that imply ages of only 2 to 5 billion years.

Further evidence that the Sculptor-type galaxies have some young stars comes from the recent discovery of carbon stars in them. The kind of luminous carbon stars found in small numbers among their thousands of bright red giant stars are believed to be fairly massive and have calculated ages (still somewhat tentative) of 2 billion years or so.

Color–magnitude diagrams of the Sculptor galaxies also are different from those of globular clusters in various subtle ways. The giant branches are thicker, either because of a spread in age or in chemical composition, or both, and the horizontal branches are, in some cases, placed differently from the expected position. This may be a result of a different *mean* age for the stars in the Sculptor galaxies. Possibly even their oldest stars are younger than the globular clusters in the Milky Way, a possibility that present techniques have difficulty testing.

Two fascinating, but still speculative, ideas have been put forward recently to explain the dwarf ellipticals and their anomalies. One is the idea that they were produced by the tidal violence that occurred 2 to 3 billion years ago when the Magellanic Clouds passed close to our galaxy. Possibly that event also disrupted the SMC into its present two components and gave birth to the Magellanic Stream. Other astronomers have pointed out that the proximity of some of the dwarf ellipticals to the Magellanic Stream suggests a common origin, though the components involved are very different: only old stars in the Sculptor galaxies and only uncondensed gas in the Stream.

A second idea, possibly related, is that the dwarf elliptical galaxies are the skeletons of small *irregular* galaxies that finally exhausted their stellar-building materials (gas and dust) 2 or 3 billion years ago. Possibly the gas was swept out of them by collision with the outer parts of the Milky Way. This interpretation of the dwarfs seems to explain their structure, which can be likened more to that of irregular galaxies than normal elliptical galaxies. On the other hand, the nonstandard structure of the objects is traditionally interpreted to be the result of the tidal force on their outer parts imposed by our gigantically bigger galaxy. On this latter interpretation, astronomers have even used the outer structure of Sculptor and its six cousins to gauge the total mass of our galaxy from its tidal influence. At present, unfortunately, the results of such an exercise are too uncertain to provide an independent measurement of our galaxy's mass, a quantity of great importance and the subject of current controversy (Chapter 5).

One last curiosity about the Sculptor galaxies may be significant. The largest and intrinsically brightest one, Fornax, has 6 globular clusters within and surrounding it. They appear normal in just about every respect, except that they are somewhat larger in diameter than globulars of our galaxy, easily understood (at least qualitatively) as a result of Fornax's much smaller gravitational field. The three that were discovered at the time of Shapley's and Baade and Hubble's exploratory studies led to the conclusion among most astronomers that the Sculptor-type objects were *not,* therefore, merely globular clusters that happened to have very low densities and large distances. What was (and is) known about galaxy formation and cluster formation seems to preclude that possibility. "Dogs have fleas, but fleas don't have fleas," is one way of putting it. The globular clusters of our galaxy are explained as very early condensations in the pregalaxy gas cloud, which separated from other early-universe density fluctuations not long after the Big Bang. The clusters' densities were high enough for stars to form in them before much star formation proceeded in the galaxy, perhaps even before the disk of the galaxy collapsed to a plane due to its rotation. The clusters, in most models of this poorly understood process, need "parent" galaxies to form near. Thus, Fornax must be a true galaxy, as its 6 fleas testify.

The Nearest Spirals

The local group contains three spiral galaxies: the Milky Way, M31, and M33. Our local galaxy is so important to astronomy that it deserves a book of its own in this series (Bok and Bok's *The Milky Way*). The other two, as the nearest spirals to ours, merit a chapter to themselves.

M31, The Great Nebula in Andromeda

Among the many objects in the northern sky visible through small telescopes, M31 is one of the most special (fig. 8.1). It is immense, spread out over more than a degree, greater than twice the diameter of the moon. Bright enough to be one of the few nebulae visible without a telescope, M31 is among the most distant objects that can be seen with the unaided eye, lying in the far reaches of the local group some 2 million light-years away. But its most spectacular feature is its sheer beauty, whether observed faintly as a mysterious patch or fuzz through binoculars or admired in all its glowing glory through a big telescope. Photographs reveal its amazing structural variety—its luminous, yellow central bulge surrounded by narrow spiral arms, studded with blue stars, clusters, and dark lanes of dust. As an example of a giant spiral galaxy, it is particularly well-endowed with all of the sorts of things that astronomers must study up close before we can understand the galaxies off in the distant universe.

Hubble's epochal study of M31, published in 1929, first proved its true nature as a galaxy. After measuring its dis-

8.1 M31, the Andromeda galaxy. The upper companion is NGC
205 (see fig. 7.10), and the smaller companion galaxy just below the
center of M31 is M32 (see fig. 7.9).

tance on the basis of Cepheid variable stars that he discovered in its spiral arms, Hubble also discovered many other features which confirmed his view that the Great Nebula must be a galaxy. For instance, by comparing many of the photographs he had taken, he found 63 novae — stars that suddenly light up to great brilliance and then, in a matter of days, fade away to invisibility. The novae of M31 became as bright as the most luminous Cepheids, some even brighter, telling Hubble that they must be similar to the novae of our galaxy (figs. 8.2 and 8.3). Hubble did not know the cause of novae, but he knew how bright they became, intrinsically, in the solar neighborhood and so could conclude that they confirmed the distance he had derived for M31 from the Cepheids.

We now understand more about what causes a nova. Novae are all apparently binaries, with one member of the pair being a collapsed star, such as a white dwarf, which has completed its normal life and is now a faint, very dense, hot "cinder." Because of this star's extremely high surface gravity, anything that might fall onto its surface will set off a reaction like an atomic bomb. The other star in a nova system is an evolving star that is growing in size because of its evolution toward the red-giant phase. It is lagging behind its companion in its evolutionary sequence, probably because it is a little less massive. As it grows in diameter, some of its outer gas layers feel the pull of the denser dwarf companion, and hydrogen streams across the gap to fall onto the companion's hot surface. An immense explosion occurs as the hydrogen, compressed by the dwarf's huge gravity, fuses into helium and releases enough energy to form a brief, spectacular nova outburst.

Almost 20 years after Hubble's paper was published, a more systematic and complete study of the novae of M31 was undertaken at Mt. Wilson. For an entire year the galaxy was photographed on every available clear night. The result was the first and only study of its kind, a complete year-long record of the novae in a galaxy. Nothing like it is possible for the Milky Way because dust hides most of it from our optical view. Other galaxies are either too small to have frequent novae or else too far away to be observed. The nova survey, therefore, like so many of the studies of M31, is important not only for understanding that galaxy but also for learning new things about galaxies in general, things that we could not learn otherwise.

For instance, an interesting and important feature of the properties of novae turned up, something that would not

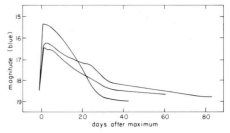

8.2 Light curves of novae in M31, based on the data published by Hubble.

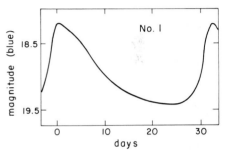

8.3 Light curve of a Cepheid in M31 (Hubble's number 1, as identified in his discovery paper).

have been so easily discovered in our local galaxy. Different novae were found to decay in brightness at different rates, and the rate of decay was found to correlate with the maximum brightness of the novae. The brightest novae decay the most rapidly. All of the novae in M31 are at relatively the same distance from us, about 2 million light-years, so we can compare them directly without having to worry about the affects of different distances, as is the case for novae in our galaxy. This important correlation, therefore, shows up more clearly for the M31 novae. If it is assumed that all novae everywhere behave similarly, then we can use the results from the M31 novae to gauge the distances to novae anywhere, providing we measure their rates of decay. The novae of nearby galaxies, through this correlation, have become a means for attacking the distance scale problem, which is discussed in Chapter 10.

S Andromeda

When the first novae in M31 were being found at Mt. Wilson, they aroused some skepticism because of their faintness. Since the true distance to Andromeda was not yet known, the faintness had no explanation, and the problem was compounded by another nova, one that had appeared in 1885. An extremely bright star, almost visible without a telescope, had appeared in that year near the center of M31. Named S Andromeda, it behaved like a nova, gradually fading until a few months later it disappeared from view. Astronomers of the time assumed that it was an ordinary nova. During the 1920s, when the nature of galaxies was being debated, some astronomers pointed to the remarkable brightness of S Andromeda as evidence that M31 cannot be very far away. However, when Hubble established the galaxy's distance on the basis of Cepheid variables, astronomers realized that S Andromeda could not be a true nova but must be another type of object, what we now call a supernova. Several of these whopping explosions had been seen long before in our own galaxy, though not recognized, and they were being discovered occasionally in more distant spirals. We now know that supernovae involve the complete destruction of a star at the end of its lifetime, sometimes leaving only a tiny neutron star as a reminder. In a giant spiral like M31, supernovae are expected to occur every 50 years or so. S Andromeda is the only one seen in M31 thus far, so we should be able to witness another of these spectacular events very soon.

The Cepheid Variables

The Cepheid variables of M31, though of great historical importance, have remained relatively neglected since Hubble detected 40 of these giant, pulsating stars and used them to gauge the galaxy's distance. About 25 years later Walter Baade and Henrietta Swope used the Palomar 200-inch telescope plates to make a more thorough study, carefully calibrated photoelectrically, of a small outer region's Cepheids, further refining the period–luminosity relation (fig. 8.4).

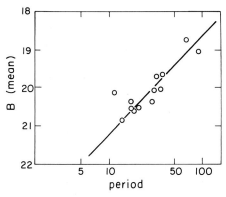

8.4 Period–luminosity diagram for M31 Cepheids as determined by Baade and Swope for a small region in the extreme southwest portion of the object.

Matter between the Stars

There is gas in Andromeda, some of it visible but most not. The visible gas appears as small, not very bright HII regions, in or near the galaxy's spiral arms. A marvelously displayed atlas of nearly 1,000 HII regions in M31 was recently compiled by a group of French astronomers at the Observatoire de Marseille. Unlike the giant gas regions of the Magellanic Clouds, those of M31 are rather small and faint. This deficiency is probably the result of the lack of massive centers of star formation, which may be lacking because conditions (neutral gas density or dynamics) are less favorable than in the Magellanic Clouds. Even the largest gas clouds in M31 are quite faint and thus an H-alpha photograph of the galaxy is unspectacular (fig. 8.5). Other Sb galaxies seem to be equally short-changed, and thus the structure must somehow be involved in the explanation. Sa galaxies are even worse off; surveys of galactic HII regions turn up only a handful of Sa's in which any HII regions were detected at all.

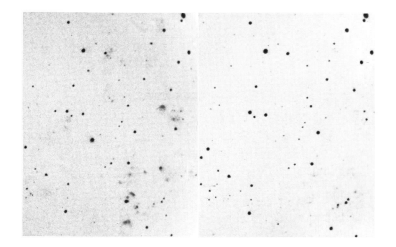

8.5 Two photographs of an outer portion of the Andromeda galaxy. The left-hand photo was taken with a filter that admits the red light of hydrogen gas (H-alpha) and the right-hand photo with a filter that excludes this light, registering only starlight. The left-hand photo shows images of several diffuse gas clouds (HII regions).

Cool, neutral hydrogen gas, on the other hand, is fairly abundant in M31, lying undisturbed in the vast voids between stars. The neutral hydrogen in M31 has been studied in great detail with large radio telescope arrays, which have produced glorious maps of the galaxy at every velocity, allowing us to trace its complex motions as well as the equally complex distribution of gas (fig. 8.6). The neutral hydrogen is not spread out in the galaxy like the stars, but shows a large gap at the middle and a concentration at about 40,000 light-years away from the nucleus. This is also where the visible gas clouds are most numerous and where the spiral structure is the most intense. Unlike the optical HII regions, which taper off at distances of about 50,000 light-years, the neutral gas as traced in radio maps extends to 100,000 light-years, far beyond most of the visible parts of the galaxy. This is three times as far as the sun is from the center of our galaxy and 50% farther out than the remotest gas in the Milky Way's plane.

The gas does not all swing around the center of M31 in perfect circles, as astronomers expected. Instead, some of it behaves quite strangely. The inner arm in the northeast part

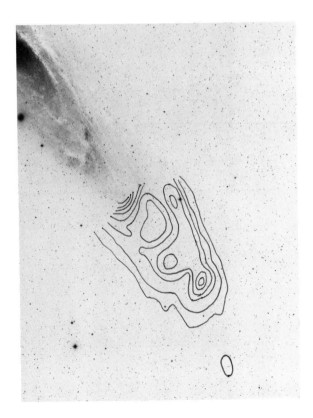

8.6 Detail of HI gas clouds in a distant outer part of M31, plotted on a negative photograph of the region. Notice the large amount of hydrogen that lies far beyond the optically bright part of the galaxy.

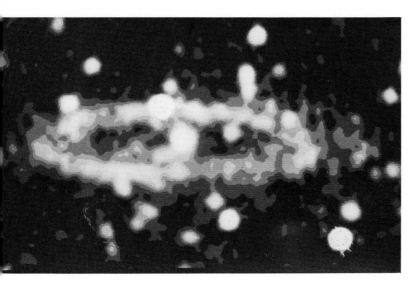

of M31 is falling in toward the center, as well as racing around in its orbit. The infall velocity is as large as 100 km per second (about 200,000 miles per hour)! We still do not know the reason for this anomalous motion; it may be due to an explosive event, some giant super-supernova, or perhaps it results from tidal action on the arm caused by one of the companion galaxies. Further analysis of this and similar non-circular motion in M31 should eventually help solve this mystery.

Maps of continuum radio radiation emitted by M31 have been made with several of the world's biggest radio telescopes (fig. 8.7). A large amount of the radio light comes from the center of the galaxy, where a very bright source is located, probably a nonthermal source related to the nuclear activity of M31. Perhaps there is the remnant of a large collapsed object at the nucleus, maybe even a black hole, though M31 is certainly not a radio galaxy in the normal sense of the word (Chapter 11). Other nonthermal radio light comes weakly from the disk of the galaxy. The intensity and character of the radio radiation is about what we would expect from the number of supernova remnants that should be there, based on a sample of objects like the Crab Nebula from our own galaxy. A few of the remnants are just bright enough to be seen and analyzed optically. The radio and optical data indicate that the supernova remnants are concentrated near the central bulge and extend thinly throughout the disk.

The rest of the radio light that is detected seems to come

from the hot HII regions in the disk, especially from the brighter parts of the spiral arms. There is a strong maximum at a distance of about 30,000 light-years, where the neutral hydrogen and star formation are also concentrated. A doughnut-shaped ring contains most of the action in M31, out about the distance from the nucleus that corresponds to the sun's distance from our nucleus.

Planetary nebulae in M31 seem to exist in areas where there are mostly old stars, especially in the central, giant-rich bulge. Astronomers have used the Lick Observatory 120-inch telescope recently to discover 315 planetary nebulae in just a small region of M31, and they estimate that the whole galaxy has about 10,000 of these tiny, ghostly objects.

Stars and Clusters

About as small and hard to detect as the planetary nebulae are the open star clusters of Andromeda (fig. 8.8). Hubble found some, apparently, since he pointed out an example in a photograph in one of his papers. But no one subsequently seems to have looked for or at them until the 1980s, when the 4-meter telescope at Kitt Peak National Observatory was used to discover over 400 of these faint, indistinct objects. Most of those that could be detected must be fairly young clusters, like the Double Cluster in Perseus (h and chi Persei), with ages less than about 100 million years. Undoubtedly there are thousands more — older clusters, whose stars are too faint to be seen and whose total brightness is low enough to be beyond the limits of the survey.

M31's spiral structure is punctuated by almost 200 large stellar associations, rather surprisingly different from those found in the solar neighborhood (fig. 8.9). They contain the same kinds of brilliant blue stars and gas clouds, but are almost 10 times as large as the local sample. Instead of being about 150 light-years across, like the well-known associations in Orion and Sagittarius in our galaxy, those in M31 average about 1,500 light-years across. We simply do not know why this difference occurs; other galaxies, such as the Magellanic Clouds, NGC 6822, and IC 1613, have stellar associations very similar in size to ours.

Globular clusters in M31 are easy to see, some even with moderate-sized telescopes (the brightest clusters are about magnitude 14). Hubble first catalogued nearly 200 of these luminous but unresolved objects; more recent efforts by independent teams of astronomers in California, Canada, Italy, and the Soviet Union have tripled Hubble's number. Globular clusters are now being used as probes of the gal-

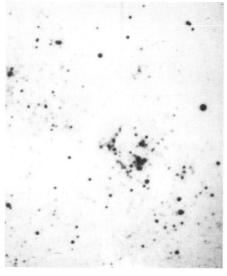

8.8 A small open star cluster in M31.

axy's old population and its history, as well as the galaxy's total mass (fig. 8.10).

Many years ago Andromeda was the first galaxy ever to have its mass estimated. As we have seen, the speed with which a star or gas cloud moves in its orbit around a galaxy depends upon the amount of mass lying inside its orbit (as well as on the size of the orbit, of course). By plotting out the velocity curve, an astronomer can measure the mass distribution in the galaxy and then sum it up to get the total mass.

For M31 the best optical data seemed to indicate that the mass must be about twice that of the Milky Way, roughly 200,000,000,000 times the mass of the sun (fig. 8.11). However, recent radio observations disagree with that conclusion, as described in Chapter 5. Most of M31 is way out where we can see nothing, far beyond the optical image and even beyond the radio limits. Other galaxies, even our own, have this same property, it seems. M31 was the first galaxy to present us with this cosmic puzzle.

Spiral Arms

The spiral arms of Andromeda are yet another enigma. Because we see the galaxy at a steep angle, its plane being only about 12 degrees from the line of sight, it is not easy to

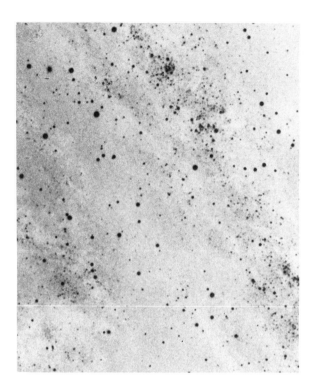

8.10 A portion of M31 containing clusters and associations.

map out just where the different parts of its structure lie (figs. 8.12 and 8.13). The structure is especially difficult to discern because there also seems to be a disturbance of the arm pattern due to the tidal effects of M32. It is possible, nevertheless, to perceive a two-armed spiral pattern with arms arranged so that they trail behind the rotation of the galaxy. Theoretical models of galaxies seem to indicate that this kind of rotation, with the arms trailing, is the most common arrangement. But more detailed attempts to disentangle the spiral structure have led to contrary results. A group of French astronomers has found evidence that M31 may have only one arm, which may be leading, rather than trailing (that is, the outer tip of the arm comes first, being the front rather than the tail of the rotating arm).

The controversy about the arms is still unresolved, but the most likely outcome is agreement that the arms of Andromeda, like those of so many galaxies, are not very regular; they do not conform to the perfection of theoretical models, but have distortions due to tidal effects and are fragmented and imperfect. The galaxy's steep angle of inclination just confounds the problem, while tempting us to hope for more regularity than we have reason to expect from any galaxy.

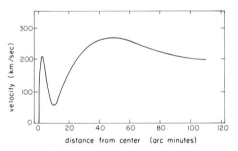

8.11 Rotation curve for M31, based on optical data. Compare with the radio rotation curve shown in Figure 5.8.

8.12 The dust arm of M31.

M33, The Triangulum Nebula

Not too far away in the sky from M31 lies another spiral galaxy, in the tiny constellation of Triangulum. An easy object for small telescopes, M33 is bright enough to be just barely visible to the unaided eye, if conditions are near perfect. It is both smaller and fainter than the Andromeda Nebula, and because it is at about the same distance from us, these differences are intrinsic as well as apparent. M33 is of a different Hubble type from M31, being a perfect example of type Sc, while M31 is Sb. It is a different luminosity-class object, too; its class is III, indicating its relative faintness. Fortunately for us, it is only moderately tilted to the plane of the sky, so that its glorious arms, brilliant gas clouds, and

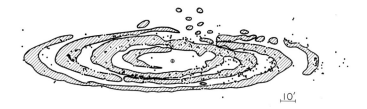

8.13 Comparison of the neutral hydrogen arms and the open cluster distribution in M31.

many bright stars are nicely laid out for us to see and study (fig. 8.14).

The HII regions of M33 are far more luminous and interesting than those of its larger neighbor (fig. 8.15). Like the Magellanic Clouds, M33 has a number of true giants among its gas clouds; three are large and bright enough to rival the great 30 Doradus. NGC 604, the biggest and brightest, contains a nest of furiously burning O stars, with temperatures as high as 50,000°K. Some are Wolf-Rayet stars, whose hot outer atmospheres boil away in the intense heat and light, producing bright emission lines that make them easy to recognize. These very massive, very young stars serve as good indicators of active star formation.

The stellar associations of M33 have been the subject of recent scrutiny. There are 143 of them, and almost all of the galaxy's most luminous stars are located here. Spectra show that the very brightest star in M33, located in one of these associations, has an absolute magnitude of −9.4, not quite as bright as the brightest stars in the Magellanic Clouds (fig.

8.14 M33, an Sc galaxy in the local group.

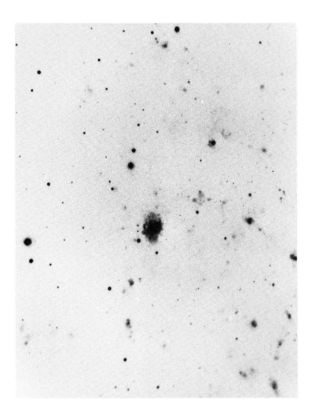

8.15 HII regions in a portion of M33. The largest is known as NGC 604 and is one of the few giant HII regions in this galaxy. This hydrogen-light photograph should be compared with Figure 8.5, which shows the much fainter type of HII regions found in the Sb galaxy M31.

8.16). Some very bright stars also occur outside the associations, despite our theoretical models which say that it is almost impossible for stars to form in isolation. Much is still to be learned about the factors that contribute to star formation, especially the formation of large, massive stars between the spiral arms.

The spiral structure of M33 is quite typical of Sc galaxies. There are two main arms, which, however, cannot be traced very far around the galaxy before they lose their distinctness. There are 10 different arm segments, and the galaxy's stellar associations, HII regions, supergiant stars, and underlying unresolved stars all are concentrated in them. The dust, on the other hand, is chaotic in its distribution and could not be used alone to discover the spiral structure. This is a strange result: What we know about spiral structure tells us that dust should be caught in the wave that traps the stars and gas. Furthermore, the dust is needed there in order to help with the difficult task of forming stars. In some galaxies, at least, the dust is well-behaved and lies conveniently along the inner side of spiral arms, but not in M33. It may be some

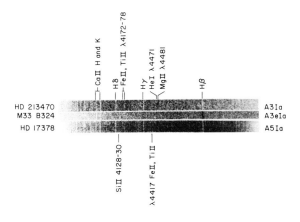

8.16 Spectrum of a supergiant star in M33 (star B324), compared with the spectra of two standard galactic supergiants of similar temperature and luminosity.

time before this interesting mystery is solved, as the dust in a galaxy is not very easy to study in the kind of detail we would like. We can map out the dust in M33 from the obscuration that it causes against the arms, but it will be very difficult to determine where it came from, how it got to its present location, what it is made of, and how it is now moving.

The star clusters of M33 have long intrigued astronomers. Hubble found a number of objects that he thought might be globular clusters, but he noted that they were significantly fainter than those in M31. More recent work shows that they also have the wrong colors, on the average. Most of Hubble's original 15 globular clusters are blue, similar in color to relatively young open clusters like the Pleiades in our galaxy. Evidentally they are actually just rich open clusters; the brightest need only be two or three times as bright as the Pleiades, which is not a particularly big cluster. The majority of the objects, then, are youngish clusters but rather large, like the blue globulars of the Magellanic Clouds. Recent, more complete surveys have isolated several hundred candidates, the majority of which are open clusters. Only a few could be genuine globulars and these are poor examples of their class, faint and small.

Why M33 should be so poor in true globular clusters, while M31 is so rich, is an important question. We can speculate that it has something to do with M33's smaller mass, but it also must be a result of its different Hubble type. Apparently globular clusters do not form in such abundance or in such large size in galaxies of type Sc or Irr I. This seems consistent with the fact that such galaxies do not have very big central bulges of old stars. They apparently did not indulge in as much star formation back near the beginning, when they first condensed out of the cosmic cloud.

The neutral hydrogen gas in M33 has been mapped and found to be rather well-behaved. It spreads out over the whole image of the galaxy, not extending too much beyond the visible disk. The velocities are fairly regular, indicating that most of the gas rotates around the nucleus in circular orbits. The total mass inside the outermost measured point is about 20 billion suns, quite a bit less than the mass of M31.

Finally, we come to the strange case of the Cepheids in M33. Hubble discovered a number of them, using their period–luminosity relation to find the galaxy's distance. Sandage recently went back over Hubble's original data, now over 50 years old and long lying dusty in a drawer, and found that he could use the measures with modern calibrations to recalculate the distance more accurately (see Chapter 10). When he found that Hubble's faint magnitudes were quite wrong, being as much as 75% too bright, he concluded that M33 must be correspondingly farther away than had been thought. In all the years since Hubble's paper, no one had reanalyzed the Cepheids; the distance to M33, which is used as a fundamental calibrator for more distant galaxies, had been based largely on very old and very shaky data. Using his new, more reliable faint standards to derive new data on even more Cepheids, especially some that are in the outer parts of M33, where photometry is not so difficult, Sandage obtained results that agree well with the revised version of Hubble's original period–luminosity diagram. Hence, the larger distance to M33 is probably correct. Such a revision of the distance would make the M33 globular clusters a little more normal in brightness. This was not to be the last word on the subject, however.

A group of astronomers at the University of Toronto have been trying a new method of measuring Cepheids that should avoid many of the troubles caused by intervening dust. Hubble had ignored dust altogether, as it was not known to be a pervasive problem back then, whereas Sandage had used multicolor photometry to try to correct for its effects. The Toronto astronomers, however, avoided much of the dust problem by observing the Cepheids in infrared light, which is much less affected by dust than is visible light. Their results for M33 disagree with Sandage's by about one magnitude, which would put M33 back at about the distance found from Hubble's original photometry. Thus the Cepheids need to be looked at again. They are essential to establishing how far away M33 is and they form one of the backbones of the entire distance scale for galaxies (see Chapter 10). Until those in M33 are unscrambled, we should not feel too surprised when our distance scales sometimes give us contradictory results.

The most recent surprise about M33 came to light only in 1983. The *Einstein* x-ray detecting satellite found an x-ray source at the nucleus of M33. Other sources had been found in the disk, probably mostly supernova remnants and binaries with a neutron star. The x-ray source at the nucleus, however, is special, similar to the x-ray sources associated with active galactic nuclei; these remarkable objects inhabit the centers of the bizarre Seyfert galaxies and quasars (Chapters 11 and 12). The nucleus of M33 is nowhere as powerful as those, being a very small example of the class, though ten times as powerful as the object inhabiting the nucleus of M31 and over 10,000 times more powerful than the source at the nucleus of our own galaxy. An unusual feature of M33 X-8, as the source is called, is the fact that it does not show up at any other wavelength. Many active galactic nuclei are strong emitters of all wavelengths of light, from x-rays to radio waves. But M33's strange nuclear feature, perhaps a black hole or other collapsed object, emits all of its power as x-rays. At these wavelengths it shines a million times more brilliantly than the sun does at all wavelengths.

Clusters and Superclusters of Galaxies

A careful look at almost any portion of the sky with a large telescope will reveal a number of faint images arranged in groups, all near the limit of visibility (fig. 9.1). These are clusters of distant galaxies. Much can be learned about the early universe and about how galaxies came into being through the study of these clusters. They are a clue to the remarkable happenings that occurred back when matter first separated from radiation, beginning its grand organization into our universe of material things.

Richness

Clusters of galaxies range in population from meagre ones like the local group to immense ones, with many tens of thousands of galaxy members, like the cluster in Coma Berenices. Astronomers refer to the number of galaxies in a cluster as its richness. How rich a cluster appears depends very much on the power of the observation. If our local group were observed from some distant galaxy, a small telescope might reveal only a loose association of three galaxies (the Milky Way, M31, and M33). Were the astronomer to take a long exposure photograph of the group, more galaxies would begin to fill in the space: NGC 205, M32, and the Magellanic Clouds would show up. But only with a very powerful telescope would there be any chance of seeing the abundant dwarfs in our group, such as Sculptor or IC 1613. Therefore, when we observe a distant cluster of galaxies, our

9.1 The Hercules cluster of galaxies, a loosely structured group containing many spirals and ellipticals. Most of the objects in this photograph are galaxies, although a few foreground stars in the Milky Way are seen in front of the cluster.

judgements of its richness are partly subjective, a function of its distance from us and the size of our telescope.

In an attempt to put all clusters on somewhat equal footing, astronomers describe the richness of a cluster in terms of the number of galaxies it contains within a certain brightness interval. For example, the great cataloger of galaxy clusters, George Abell, determined the population of clusters according to the rule that one counted those galaxies within the two-magnitude interval fainter than the third brightest galaxy. He chose the third rather than the first to avoid problems that might be caused if the first (or second) brightest might happen to be a bright foreground object; distances were not known for most galaxies when he was making his pioneering catalog, which was based on the Palomar Schmidt atlas plates that he examined as they were being taken in the 1950s. By Abell's criterion, galaxy clusters in his catalogue (which excluded small groups like the local group) range in richness from about 50 to 300. The total number of galaxies in these clusters, of course, is far larger than this, many thousands for most of the clusters Abell studied. The richness measure just skims the cream off the top.

Types of Clusters

Richness is just one way in which galaxy clusters differ. Abell also noted that some clusters seemed much more regular than others, and he separated them into two classes according to the regularity of their shape. In recent years a number of different classification systems have been invented, but all basically recognize that a cluster's types (spiral, elliptical, irregular) seem to be related to its shape. Those with a smooth, regular shape are mostly made up of elliptical and S0 galaxies, while loose, irregular clusters contain many spirals and irregular galaxies.

The Coma cluster is a spectacular example of a regular cluster (fig. 9.2). It is compact and concentrated to its center, where there are some very luminous giant galaxies, NGC 4889 being the brightest, at 13th magnitude. This elliptical galaxy is almost large and luminous enough to be called a cD galaxy, a classification reserved for certain extremely dominant giant galaxies that often inhabit the centers of regular clusters and that are often found to be radio sources. We'll

9.2 The Coma cluster of galaxies. The arrow indicates the brightest member, a giant elliptical.

return to these objects and their remarkable story a little later in this chapter.

The Coma cluster has very few spiral galaxy members, and most of these are located in the outskirts of the cluster. The central concentration of ellipticals is high, with the density falling off steeply away from the center. Like many regular clusters, Coma emits radio radiation. Some clusters of its type are remarkably strong emitters of radio waves, originating from a few large, active galaxies as well as from a general background of intergalactic gas. We can even "see" this gas in some clusters: as radio galaxies plow through the gas, they leave jets bent around behind them, like tails trailing in the wind.

About one-third of all regular clusters emit x-rays, a further indication of their hot intergalactic gas (fig. 9.3). In this case, the x-rays come from gas with a temperature of about 100,000,000°K. The best explanation of the x-ray-emitting gas is that it results from violent collisions between galaxies,

9.3 X-ray map of the cluster of galaxies Abell 85, which is dominated by a giant cD galaxy in its center.

which sweep out each other's interstellar gas, leaving them looking like gasless S0 galaxies. The gas from each is heated by the collision and left near the place of collision, to mix together with other hot gas from other collisions in the cluster to form a hot intercluster medium.

The density of regular clusters is high enough, especially near their centers, that such collisions can happen, though not everyone agrees that it can happen often enough to explain the concentration of S0 galaxies in regular clusters. Walter Baade and Lyman Spitzer first proposed this idea back in 1951, when the distance scale was thought to be much smaller than it is now. With the new scale (see Chapter 10), the galaxies in a cluster are farther apart and can collide less frequently than the calculations originally indicated. Perhaps the collisions mostly occurred at an early stage of the universe, when galaxies were just forming and the universe was smaller. In any case, it is true that an unusually large number of S0 galaxies are found near the centers of regular clusters, with a corresponding lack of spirals.

Irregular clusters, on the other hand, have lots of spiral galaxies. These clusters have an open structure, with little central concentration. Only about a quarter of them emit radio waves and less than 10% are x-ray sources. Collisions between members are expected to be rare enough that few S0 galaxies are created by the sweeping mechanism that is thought to form them in regular clusters. There is less hot gas as well, and these clusters are therefore rather quiet at radio and x-ray wavelengths.

The Virgo cluster is a good example of an irregular one (fig. 9.4). Spread out over several degrees in the sky, it contains many large, glorious spiral members, such as M100 and M61 (fig. 9.5). It is so little concentrated that its center is hard to locate without carefully counting members. There are subclusters within it and many small groups surrounding it out to great distances. Our local group is apparently an outlying suburb of the greater Virgo system.

To continue the urban analogy, the different kinds of galaxy clusters can be thought of as similar to different kinds of cities. Regular clusters are like New York, with a strong concentration to the central city, Manhattan, where one finds many giant buildings, all looking more or less alike. Irregular galaxies are more like Los Angeles, sprawled over the countryside with little central concentration. The middle area does contain a few large buildings of a variety of shapes, but there are other places miles away that also have local population centers. It is difficult to decide exactly

9.4 The central part of the Virgo cluster of galaxies. Both giant ellipticals and spirals, almost as bright, are present.

9.5 Two spiral galaxies in the Virgo cluster.

where it ends; it just slowly peters out at great distances, where there are still a few concentrations that may belong to the central city, though people will argue the point (fig. 9.6).

We live in a poor, irregular group of galaxies whose members are not especially obvious to us when we look out at the depths of space. What would it be like to live inside one of the great clusters like Coma? The density of galaxies near the middle of one of these is quite high, about 3,000 galaxies per cubic megaparsec for the top 8 magnitudes (a megaparsec, 1,000,000 parsecs, is 3.26 million light-years, so a cubic megaparsec is 35 million cubic light-years). The top 8 magnitudes includes all the giant galaxies and extends in brightness down to galaxies as faint as NGC 6822, but would not include extreme dwarfs like Sculptor. The galaxies would be packed together with an average distance of about 150,000 light-years, roughly the distance between the Milky Way and the Magellanic Clouds. Were we in such a cluster, the night sky would be filled with hundreds of galaxies bright enough to be seen easily without a telescope. It would be a magnificent sight.

9.6 Three galaxies belonging to the NGC 5566 group, a small cluster. NGC 5566 is the giant spiral on the right.

Cannibalism

We have seen that S0 galaxies in clusters are thought to be the result of collisions between two large spiral galaxies that pass through one another. What happens when instead a small galaxy encounters a much larger galaxy? If the velocity of the encounter is not too great, the more massive galaxy can simply swallow up the other one, incorporating its stars into the larger system. This process is at present the best explanation for the giant cD galaxies found near the centers of some regular clusters. A galaxy that "eats" other galaxies is called a cannibal galaxy. It would be expected to have an extended outer envelope, as cD galaxies do, and sometimes to have more than one nucleus, owing to incomplete digestion of its recent meals, as some cD galaxies do. It will be overweight, having swallowed other galaxies until its mass is greater than the mass of more normal galaxies, and it will reside near the center of a cluster, being its most massive member and thus dominating the motions of other galaxies. The violence of the meals should result in a hot plasma of gas visible as radio radiation. All these characteristics are true for cD galaxies, and the evidence seems overwhelming that they are the result of this process. Fortunately, cannibalistic galaxies are rare, occurring only in the dense centers of clusters much richer than our local group. We are spared the embarrassment of becoming either an eater or an eaten.

Superclusters

Several early students of galaxy clusters noticed a tendency of some clusters to be clustered together in even grander clusters. One impressive example is the way that nearby groups seem to congregate around the Virgo cluster, making up what has come to be called the local supercluster. When Abell examined his catalog of clusters, he found evidence for this kind of superclustering. The groups of clusters that he found seemed to be about 100 million light-years across and included an average of 6 clusters.

Recent years have corroborated the existence of superclustering. Various kinds of statistical tests are necessary to be sure of the reality of such clumps; it is remarkably easy to see clustering that is not there, so astronomers have exercised great care in this respect. Most tests indicate that clusters of galaxies are frequently clustered together in groups ranging in diameter from about 100 to about 1,000 million light-years. They have masses about 10^{16} times the mass of the sun. Most are not round in outline, but long and narrow.

The local supercluster is the only one found that has a significant concentration to its center, which is in the region of the Virgo cluster. The other well-studied examples seem to have no particularly dense center, but are just loose groupings of roughly equal clusters (fig. 9.7). The local supercluster's center is about 60 million light-years from us and the local supercluster's diameter is about 120 million light-years. It is flat, with the ratio of its major and minor axes being approximately 6 to 1.

A much larger example is the Corona Borealis supercluster. It is one and a half billion light-years from us and has a diameter of over a billion light-years, which means that it occupies much of the space that we can see when we look in the direction of that constellation.

Superclusters are rather different kinds of things from star clusters or clusters of galaxies. For one thing, they are not dynamically well-defined, as there has not been enough time in the life of the universe for them to respond to each other's gravitational pull. The time it would take one cluster to cross the supercluster is typically about 300 billion years, many times the age of the universe. This easily explains why superclusters do not have a regular structure and are usually

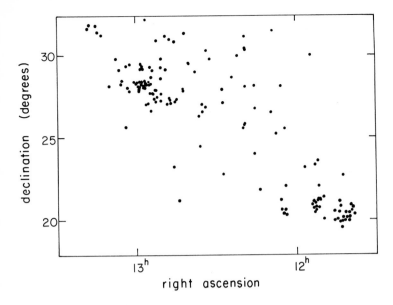

9.7 The positions in the sky of the galaxies in the Coma-A1367 supercluster. All galaxies in this map are at about the same distance from us, some 300 million light years.

not concentrated to their centers, both conditions that result from gravitational action and mixing.

The origin of superclusters is an intriguing puzzle. Some evidence seems to suggest that they must have been in the works long before the galaxies themselves formed. Theories of how they might have arisen from fluctuations in the density of matter in the early universe have been proposed. Alternatively, there is reason to believe that superclustering resulted from "small" fluctuations in space curvature, where the density of both radiation and matter were perturbed. The most exciting and exacting challenges in current cosmological research involve the question of the origin of clusters and superclusters of galaxies.

Voids

In 1981 newspapers announced the discovery of an immense area in space as large as a supercluster but almost devoid of galaxies, either separate or in clusters. The discoverers called this a "void" and pointed out that cosmologists must be able to explain the lack of galaxies as well as the presence of them. Several other voids have now been identified, the largest of which is 2 billion by 1 billion light-years in size. Along with these discoveries has come the realization that galaxies are not necessarily best thought of as objects that sometimes cluster. Instead, at least in some parts of the universe, they seem to be arranged in a network, with great empty spaces between them. The two concepts are like a positive versus a negative. Do galaxies exist in bunches embedded in a matrix of nothing or are there bunches of nothing embedded in a matrix of galaxies? Theorists rather like the latter picture, which is predicted by some models of the early universe. The universe, they say, may be like a house; the matter is primarily located in a network of walls, floors, and ceilings, but it is the empty rooms that make the house a home.

The Distance Scale

10

Over mere millions of kilometers, astronomy has found methods of measuring distances remarkably reliably, to better than one part in 10^6. But for the expanses of extragalactic space, where the distances are measured in terms of 10^{20} to 10^{24} km, the methods are strained and the accuracy is low. Considering the immensity of the job, we should feel fortunate that we are now able to argue about only a 50 percent uncertainty. But the nature of science is such that we want to know these distances much more accurately than that. Heroic efforts to push techniques as far as possible have resulted in a remarkable situation, in which protagonists have lined up behind two very different approaches to the problem and two distinctly different distance scales.

The Hubble Distance Scale

Starting in the 1920s with his pioneering work on galaxies in the local group, Hubble began an elaborate campaign to set up a distance scale that would extend to the edge of the observable universe (fig. 10.1). At first the scheme was crude, depending far too much on the assumption of uniformity. But it did lead to the first real appreciation of the vastness of the cosmos, and its method has been followed, in principle, by most astronomers of succeeding generations.

Hubble's first task was to measure distances to members of the local group. He emphasized the galaxies M31, M33, and NGC 6822, in all of which he had discovered Cepheids

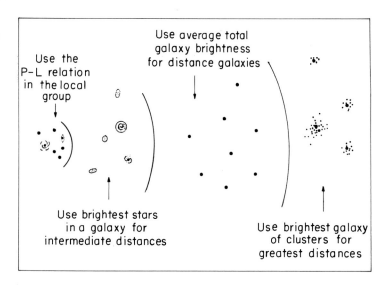

Use the
P–L relation
in the local
group

Use average total
galaxy brightness
for distance galaxies

Use brightest stars
in a galaxy for
intermediate distances

Use brightest galaxy
of clusters for
greatest distances

10.1 The Hubble plan for measuring distances out to the faintest, most distant galaxies.

whose period–luminosity relation he could use. His results for these three galaxies, combined with Shapley's data on the Magellanic Clouds and Baade's perusal of resolved galaxies in the local group, formed the baseline and first step in Hubble's three-step cosmic distance scale. We now have better measures of the distances to these galaxies, because of a revised period–luminosity relation for Cepheids and improved brightness measures of faint stars, and all are two to three times as far as Hubble first thought. But distances to the galaxies of the local group still form the framework upon which almost all distance scales are built.

Hubble's plan was, next, to use the nearby galaxies and their distances to calibrate luminosities of objects brighter than the Cepheid variables, so as to measure distances into farther reaches of space where even the 100-inch telescope could not detect Cepheids. After trying several kinds of objects, including red giants, star clusters, HII regions, novae, and so forth, he found that the brightest stars in a galaxy all seemed to reach a maximum luminosity that was fairly uniform from galaxy to galaxy. Therefore, the apparent luminosity of a galaxy's brightest stars would be related to the galaxy's distance from the viewer. A large collection of plates of many galaxies whose brightest stars were resolvable eventually provided Hubble with data that seemed to demonstrate the validity of this approach. He compiled measures of luminosities of the brightest stars in a long list of galaxies, and, as step 2, calibrated their distances by comparing these luminosities to those of the brightest stars in galaxies of the local group, whose distances were known from the Cepheid data. Finally, Hubble's list of resolvable galaxies provided him with information on the intrinsic *total* luminosities of galaxies and their dispersion. Step 3, then, applied these luminosities to even more distant galaxies, beyond the range where individual stars could be resolved. This last step could be extended to the edge of the visible universe, and it led to Hubble's most ambitious and grandest project — determining the size and the shape of the entire cosmos.

During this period, Hubble, V. M. Slipher, Milton Humason, and others had been taking spectra of galaxies and finding that some galaxies were moving with remarkable velocities, as measured by the Doppler shift of their spectral lines.* In 1928, the cosmologist H. P. Robertson pointed out a correlation between the velocities and brightnesses of galaxies, whereby the velocity of distant galaxies is greater for the fainter objects. At about the same time, Hubble showed that this indicates an expanding universe, with the velocity between galaxies being directly proportional to their separation. Almost all galaxies showed red shifts, indicating velocities away from us; only a few galaxies in our local group and a few members of nearby clusters had blue shifts. The Andromeda galaxy, for example, was found to have a 300 km/sec velocity toward us, but that is partly due to its membership in a gravitationally interacting group (perhaps it and we are in quasi-orbital motion) and partly due to our sun's own orbital motion in the Milky Way. But more distant galaxies are all receding; for instance, among the galaxies of the cluster in the constellation Virgo the average velocity of recession is 1,000 km/sec. To date astronomers have detected objects with recession velocities of at least 80% of the velocity of light. The relationship between the velocities and distances of galaxies is referred to as the *Hubble relation,* and the constant of proportionality is called the *Hubble constant,* abbreviated as H (fig. 10.2). Current workers usually refer to it as H_0, the subscript referring to the present value, as it may have had a different value in the past (which is, of course, what we see at large distances), depending on whether or not the expansion of the universe has been constant or has accelerated or decelerated. In Hubble's final years the 200-inch telescope at Palomar Mountain was fin-

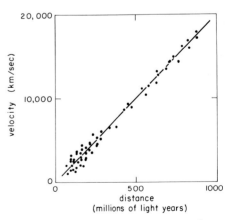

10.2 The Hubble relation between the velocity of recession of galaxies and their apparent brightnesses.

* The Doppler effect is a change of perceived wavelength in light from an object that is moving either toward or away from an observer. A velocity of approach causes a blue shift and one of recession a red shift.

ished and he began the task of using this powerful instrument to probe even farther into space and back into the past.

The Impact of Palomar

A milestone along the path toward a reliable cosmic distance scale came in 1958, when Allan Sandage, on whose shoulders Hubble's mantle had fallen five years before, gave a lecture on the occasion of his accepting the Warner Prize of the American Astronomical Society, awarded for outstanding achievement on the part of a young astronomer. The prize had resulted from his pioneering work in stellar evolution, but the lecture was on the distance scale and it caused something of a sensation. He showed some of the first results on this problem from the Palomar telescope. Reworking Hubble's original sample of galaxies using the new big telescope and more refined techniques, Sandage found several large errors in the early work, expecially in the identification of a galaxies' brightest stars. In 1936 Hubble had warned that identification of the brightest stars in a galaxy was not a trivial problem; star clusters, star clouds, and HII regions can all look stellar if viewed from far enough away. Sandage's startling conclusion in 1958 was that this problem had actually greatly affected Hubble's distances. Combined with earlier corrections, Sandage's new findings yielded a distance scale that was seven times larger than Hubble's 1936 scale. The Virgo cluster, for example, which Hubble measured to be 7 million light-years away, was found by Sandage to be 50 million light-years distant. The whole universe was vastly larger than had been thought.

Sandage has continued to pursue this problem with continually increasing care, making each step more secure. In 1974 he and his colleagues began a series of papers that developed the problem once more, this time with several new intermediate steps that bolstered some of the weakest places in Hubble's bold plan. They were able to extend the use of Cepheid variables beyond the local group by training the 200-inch telescope on Cepheids in the spiral galaxy NGC 2403, one of a small group of galaxies 10 million light-years distant (figs. 10.3 and 10.4). With two groups of calibrating galaxies (11 objects in all), they developed an auxiliary method that allowed them to stride out into the awkward region that they called the "twilight zone," between 50 and 300 million light-years. Here is where Hubble's work had gone wrong, where he had mistaken other kinds of objects for brightest stars. Sandage's approach used the apparent

10.3 The largest HII regions in NGC 2403 show up as diffuse clouds in this photograph taken in hydrogen light.

10.4 HoII, a galaxy in the NGC 2403 group, one of the nearby small clusters that is used to calibrate the Hubble constant.

sizes of a galaxy's largest HII regions as the new indicator of distance, because HII regions are resolvable out to the Virgo cluster and beyond. They provided an independent criterion, somewhat less subject to systematic identification problems than the brightest-star criterion. With it, the astronomers added a third group of nearby galaxies, centered on the giant spiral galaxy M101, to their calibrating base.

The Virgo Cluster Problem

In the meantime, the Virgo cluster had fallen into disfavor as a source of information on the distance scale and Hubble's constant, for reasons that are quite remarkable. Although several astronomers had been arguing for years that we might be an outlying part of the Virgo cluster, convincing evidence did not appear until the middle 1970s. By then enough velocities of galaxies had been measured throughout the sky, expecially as the result of a massive project carried out at the Harvard-Smithsonian-Whipple Observatories, that the anisotropy* of galaxy motions could be plotted and understood. Although still clouded by controversy, the evidence now looks convincing that our galaxy and the local group are falling in toward the center of the huge supercluster centered near the Virgo cluster. Because our local group is gravitationally involved with this supercluster, the measured velocities of the Virgo galaxies will not give a true value of the cosmic expansion velocity, as this other motion is also involved. The fact is itself an important clue to the question of how galaxy clusters form and how the mass of the universe was organized into its present complexity. But for the purposes of determining cosmic distances, it is a complication (fig. 10.5).

One method used to check on this problem was to observe the properties, particularly the luminosities and colors, of a large number of galaxies inside and outside of clusters. A comparison of field galaxies with Virgo cluster galaxies shows that the local group must be falling into the center of the Virgo supercluster with a velocity of about 290 km/sec. Other combinations of distances and motions of galaxies have been used to derive a value of 255 km/sec for the motion of the local group toward a similar position. The cosmic background radiation, discovered in 1965 and identified as the relic radiation from the beginning of the uni-

---

* Antisotropy measures the departure from an isotropic universe, in which the velocities are exactly the same in all directions.

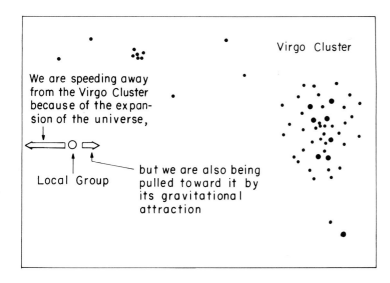

10.5 The Virgo cluster problem: to disentangle the general cosmic expansion velocity from any local velocity resulting from our local group's motion toward the Virgo cluster, in response to its gravitational pull.

verse, also shows an anistropy in roughly the same direction and of the same amount.

Luminosity Effects

An important element in the final step on the path toward a cosmic distance scale is the luminosity classification of galaxies (Chapter 1). Sandage and his colleagues used it, after calibrating it in nearby groups, and determined distances for 60 distant high-luminosity galaxies (of class ScI) with velocities ranging from 3,000 to 15,500 km/sec. Comparing the distances with the velocities gave them an answer: the Hubble constant is even smaller (and thus, the universe is larger) than previously thought. While Hubble had derived a value of H_0 of 160 km/sec/million light-years, which Sandage had reduced in 1958 to 23, Sandage now quoted a value of only 15, with an estimated uncertainty of 10%.

The Two Contrasting Approaches

The story of the development of the modern scale of cosmic distances has been recounted so far primarily from the point of view of Hubble and his intellectual descendants. A complete history of the problem, however, would have included a number of other astronomers who have done a variety of new and original things in the quest of better distances. Some reach a final answer quite different from that of Sandage.

Gerard de Vaucouleurs has long involved himself in measuring and assembling data on galaxies. Having tried a variety of different methods to measure distances over his many years of activity, finally, in 1978, he carried out an exhaustive study of the distance scale from a different point of view, rejecting the principles and greatly altering the practices of Hubble and Sandage. Instead of basing distances in the various regimes on one or two "best" or most reliable techniques, de Vaucouleurs argued for the use of a variety of different indicators, based on the argument that if one or another is faulty, the use of many will dilute the effect. Using this statistical approach, he gathered a great deal of data from the literature and treated it in elaborate detail. Whereas Sandage based his value of H_0 primarily on 5 distance indicators, de Vaucouleurs introduced 13.

The two scales begin to diverge right in our own galaxy, where different approaches to the problem of absorption by interstellar dust lead to different distances for even the nearest galaxies (fig. 10.6). Sandage determined dust absorption in our galaxy by measuring (or adopting other people's measures of) the amount of reddening it causes in front of each

10.6 Dust in the Milky Way shows up by the way it obscures more distant stars in photographs of star fields like this.

calibrating galaxy. This led Sandage and his co-workers to find very small amounts of absorption in the direction of the poles of our galaxy (latitude 90°), and to assume zero absorption for galactic latitudes higher than 50°. For de Vaucouleurs, however, the problem was very different. He assembled the mean colors of all measured background galaxies, segregated them by morphological type and inclination angle, and solved statistically for an equation that relates their corrected colors to their latitude and longitude. He similarly examined counts of galaxies, and compared their hydrogen content (measured at radio wavelengths) to their optical luminosities. Analyzing these data for an equation that describes the effects of galactic absorption, then combining the results with those from galaxy colors, he derived a statistically based equation that was then applied to all galaxies, including the nearby calibrating ones, to establish their true, unabsorbed, and unreddened colors.

The result at large distances where local effects are no longer important is a scale almost exactly one-half as large as Sandage's. That is, de Vaucouleurs' universe is half the size, and his Hubble constant is twice as large, as the one Sandage found.

Some Recent Advances

In the last few years, during the time that the two contrasting approaches were diverging, more astronomers entered the field. One of the most important new developments was the introduction of the HI line width method, using what is called the Tully–Fisher relation (fig. 10.7). This method involves a remarkably tight correlation between the neutral hydrogen gas (HI) line width, which measures the overall dispersion of velocities in the gas, and the absolute (intrinsic) luminosity of a galaxy. Although an empirical discovery, it is explainable in terms of the relation between the mass and mass distribution in a galaxy and its total number of stars. The Tully–Fisher relation subsequently has been employed by the two opposing camps — each finding results that agree with their own very different distance scales! This remarkable situation can be traced to the fact that the ultimate answer rests on the nearby calibration of Andromeda and M33 and on the treatment of internal absorption in the galaxies. These two problems are handled differently enough to produce a complete dichotomy of answers.

In 1980 a new treatment of the Tully–Fisher relation emerged that was designed to avoid much of the uncertainty

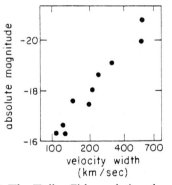

10.7 The Tully–Fisher relation shows that galaxies with wider velocity widths are more luminous.

in the question of internal absorption of the sample galaxies. Because dust absorbs relatively little infrared radiation, total infrared luminosities are more certain than visible-light luminosities. A recent series of studies using infrared luminosities of distant galaxies in clusters gave results that are rather closer to the short distance scale (de Vaucouleurs') than to the long one, even though Sandage's distances were used for the calibrating galaxies and his distances for the nearest groups were found to agree with the infrared studies.

A related attack on the same problem recently used detailed velocity curves, which give the full picture of how stars revolve in the spiral disk of a galaxy, to obtain a distance scale. The results of this analysis so far fall about midway between the two opposing scales, and agree, in fact, rather well with Sandage's 1958 result, which said that the Hubble constant is about 23 km/sec/million light-years.

Even more refined looks at various steps in the distance scale are now being made: the HII-diameter method has been refined, the globular-cluster-luminosity method has been exploited, and the brightest-red-supergiant criterion has been recalibrated. All of these have helped to clarify a confusing situation, but not to solve the problem. All lead to values of the Hubble constant in the range spanned by the two opposing scales, but none definitively settles what the

10.8 A very distant cluster of galaxies is just barely visible in this deep photograph. It illustrates the extremely difficult job astronomers face in trying to study the most distant galaxies.

right value is. The diversity of these answers tells us clearly that the fundamental data are both less accurate and less well understood than assumed by their users.

Since Hubble first proved that galaxies span incredible stretches of space and time, some of the boldest minds in astronomy have attempted to gauge the extent of the universe. Rather than being disturbed and discouraged that the scale of distances is still a matter of dispute, we should probably be grateful that nature has made it possible for us to find methods to measure a cosmos so unimaginably big.

Disturbed
Galaxies

11

$\mathbf{I}$n the early days of radio astronomy it was not easy to figure out what was emitting the radiation that was being detected. Because radio waves are so long, typically about the length of this book, images received by radio telescopes are very blurred. The resolution of any kind of telescope depends on the ratio of the wavelength to the telescope aperture; so to improve the resolution at a given wavelength, one must increase the telescope size (figs. 11.1–11.3). But even giant radio telescopes, much larger than the largest optical telescopes, can have poorer resolution than even the smallest of optical telescopes. Therefore, in the 1950s, when radio telescopes were still relatively small single dishes, it was difficult to see sources clearly and to be sure of their exact positions. Radio astronomers knew roughly where their objects were; they could at least say in which constellation they resided. But without more accurate positions it was difficult to figure out just which objects were giving out such surprisingly strong radio signals.

The First Discoveries

The first catalogued radio sources were named after their constellations. The brightest was assigned the letter A, the next brightest B, and so on. Thus, the brightest radio object in the sky was named Cassiopeia A, or Cas A for short. Other early discoveries were Cygnus A (Cyg A) and Centaurus A (Cen A).

The first indication that some of these radio sources might be distant galaxies came in 1949, when Australian astronomers identified the strong radio source known as Cen A with the peculiar galaxy NGC 5128. The question was settled in 1954, when Palomar Observatory astronomers used the 200-inch telescope to search for visible objects that might coincide with other bright radio sources. After considerable effort, they found several peculiar objects that were good candidates and tried to figure out what they might be. One of them, a faint, strangely-shaped image at the position of Cygnus A, turned out to have a spectrum like that of a distant galaxy, whose redshift indicated a distance of about 700 million light-years. Other candidates were also galaxies with peculiar structure of one sort or another. Astronomers concluded that galaxies can be strong radio sources—but only very strange galaxies.

Galaxies in Collision?

Some of the first radio sources looked like two galaxies that might be colliding, and it seemed that galactic collisions might be the violent cause of the tremendously energetic event implied by the radio noise. But other radio galaxies showed no signs of collision. And besides, it was clear that simple encounters between galaxies would not necessarily lead to a violent reaction that would result in a strong radio source (fig. 11.4). The stars in a galaxy are so far apart compared with their diameters that a galaxy is nearly empty, as far as stars are concerned. The collision of two galaxies is not likely to involve a single collision of two stars. The galaxies will merely merge and then reemerge as they pass through each other. At first glance, therefore, it seems unlikely that a collision of galaxies would be a very interesting event, certainly not a spectacular one that would generate a strong radio signal.

But three possible kinds of events can occur that might make a big difference. (1) A collision can produce violent tides that tear the galaxies apart; (2) the collision of gas and dust in the two galaxies can lead to a hot, shocked medium that can radiate its alarm; and (3) in the case of galaxies of quite different masses, the larger one can swallow the smaller. For each of these kinds of events there are suspected candidates; some will be described below. But in the years immediately following the apparent discovery of colliding galaxies, the whole idea died a sudden and, as it turned out, undeserved death.

11.1 A radio telescope, such as this one of the Harvard-Smithsonian Center for Astrophysics, looks something like a television receiver, except that it can turn to any point in the sky.

11.2 The five-kilometer radio interferometer of Cambridge University.

11.3 The interferometer at Westerbork in Holland.

11.4 NGC 520, two galaxies that are probably merging, though they are not a radio source.

As more radio galaxies became identified, some of them turned out to be so bright that there was no obvious way that enough energy could be generated by a collision (fig. 11.5). Others consisted of a single galaxy (fig. 11.6) and often the solitary object was a nearly gasless elliptical. As it became clear that something besides collision had to be thought up to explain most radio sources, other hypotheses began to flow from the pens of theorists. Among them were models that involved antimatter galaxies, magnetic flares, accretion of intergalactic matter, formation of new galaxies, chain-reaction supernovae, formation of new matter, and the action of a central, supermassive object. In the years since, this list has been pared down. New telescopes, new techniques, and thousands of new radio galaxies gradually caused the elimination of almost all the hypotheses. At present, though we still do not really know what causes radio galaxies, the choices have been narrowed down to just about one possibility, the only one that seems capable of generating such immense amounts of energy and causing the many weird shapes and strange features of the radio galaxies.

In this model, a massive object, probably a black hole, resides at the galaxy's center, causing havoc as it inexorably sucks in the matter around it, matter whose final screams we hear in the radio spectrum. In order not to prejudge the case, however, astronomers do not usually speak of this thing as a black hole, but use some less specific term, such as "supermassive object." Some astronomers, anxious to preserve their neutrality but not to obscure the necessarily remarkable nature of this object, call it the "central monster."

11.5 The arrow in this figure points to 3C 31, a radio galaxy in a small cluster of galaxies.

The Monster at the Center

Radio telescopes show that there is a tiny, central, powerful object in most "active galaxies" (those that emit radio or x-ray radiation or both). Intercontinental interferometers have demonstrated that these central objects must be very small indeed, less than about 1/10 of a light-year across. They seem to be the source of great jets of material and radiation, usually emitted in two opposing directions. In some galaxies these jets extend outward from the nucleus in a compact double formation, detectable at optical, radio, and x-ray wavelengths. In others, the jets extend far beyond the visible galaxy, reaching out into the intergalactic space (fig. 11.7).

The radiation making up the jets is characteristic of the kind emitted when electrons are moving at near-to-light

11.6 M82, a remarkable radio galaxy in a nearby group.

speeds through a strong magnetic field. This light is called "synchrotron radiation" because it is seen by physicists as a glow in high-energy particle accelerators, such as synchrotrons. When the light of radio galaxies was first found to glow with these same characteristics, it was proposed that the radio radiation might be explained in this way, and now the evidence is overwhelmingly in favor of this idea. The jets that are seen in such galaxies as M87 (Virgo A) and NGC 5128 (Cen A) show up both in visible light and at radio wavelengths because they are the paths of extremely high-

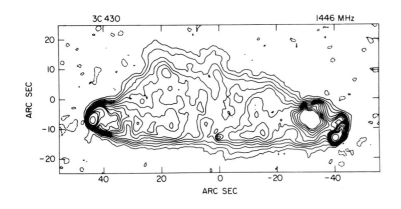

11.7 A radio map of the radio galaxy 3C 430, made with the VLA. The optically visible galaxy is located midway between the two immense radio "lobes," which map out the location of high-energy particles that were ejected violently from the active radio galaxy at the center.

energy particles (mostly electrons) that are hurtling through the galaxy along with a strong magnetic field.

What is this thing in the center that can cause such spectacular fireworks? As mentioned above, the best explanation is that it is a supermassive object, probably a black hole. Black holes result from the complete collapse of a star or cloud that lacks enough rotation or internal energy to withstand the gravitational force of its own enormous mass. General relativity tells us that such an object will disappear from our view when it reaches a certain stage in its collapse, and that from then on no radiation can ever escape from it. Anything that falls into it will never be able to return. The density soon becomes incredibly large and strange relativistic effects commence, making little sense to our earth-bound minds. But the important point here is that a rapidly rotating "accretion" disk forms around the black hole, made of the gas that is being pulled into it. Before this material disappears down the hole, it has so much energy that it can squirt jets of relativistic electrons out from its poles, and these produce the radio patterns we see in radio galaxies (fig. 11.8).

Food for the Monster

To account for the tremendous outlay of energy in a radio galaxy, some form of fuel must be found. Where does the gas come from that falls into the center, leading to the accretion

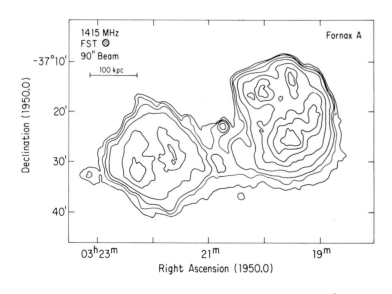

11.8 Map of the radio radiation from the nearby radio galaxy Fornax A, a disturbed, dusty elliptical galaxy in the Fornax cluster. The optical galaxy is located at the center of this radio map.

disk and the jets? Calculations indicate that at least one sun's mass per year is necessary. What stars or other things are being sacrificed at this rate? In the ordinary course of events stars do not fall into the center of a galaxy no matter what is in there. They continue to revolve around it in their orbits, unless acted upon by some extraneous force. In the early days of the galaxy, of course, there would have been some left-over gas in the disk that could fall to the center, not having the right amount of angular momentum to maintain an orbit. But this stuff should be all gone by now; instead, we must find a source that can be putting matter into the nucleus now, or at least very recently (to astronomers that means within the last billion years).

One source that looked reasonable at first was the matter lost regularly by stars as they evolve. We know that this process continually puts stellar material back into the interstellar gas so that it becomes enriched with heavier elements. Calculations of the rate of loss of stellar gas, however, indicate that only about 10% or less of the necessary quantity of gas could be acquired by the monster from this source. To satisfy its amazing appetite, we must look elsewhere for a provisioner.

A more promising possibility is galactic cannibalism. If a little galaxy should get too close to a big galaxy, the gas, dust, and stars in the disk of the smaller object would be tidally stripped away from its nucleus early on in the drama and would melt into the outer parts of the larger galaxy. The central nucleus of the smaller galaxy would not be easily disrupted, however, and should fall down to the center of the big galaxy. This source would easily provide the monster with enough full meals to keep up its activity for millions of years. Such collisions would not happen often, of course. But radio galaxies are fairly rare; only about 1% of the galaxies around us are markedly disturbed. Not too many close encounters would be needed to explain this rate. Moreover, radio galaxies are most common in galaxy clusters, where the chance of collision is greatest. More about the strange beast in the centers of active galaxies and about the possibility that galaxies eat each other is given in Chapter 12.

AGNs and Other Oddities

The objects discussed so far in this chapter have all been referred to as radio galaxies. As our knowledge of galaxies has progressed, we have seen many features common to a number of different kinds of unusual galaxies, many of which are not especially strong radio emitters. For example, in the 1940s Carl Seyfert discovered a class of galaxies that have extremely bright and broad emission lines in their nuclei. These are now called Seyfert galaxies and are the subject of many enlightening studies. An example is the spiral galaxy NGC 1068, which is not only a Seyfert galaxy but also a radio source (fig. 11.9). In its nucleus is a very bright, hot gas cloud that shows turbulent velocities of thousands of kilometers per second. In energy it resembles other radio galaxies, and seems to be the weak end of a spectrum of similar objects, ranging up into the strong radio galaxies and the quasars (Chapter 12). All of these objects have in common a definite, very energetic disturbance of some kind in their nuclei. Frequently astronomers refer to these violently disturbed central regions as AGNs, for active galactic nuclei.

Another class of peculiar objects, which look something like Seyfert galaxies, are the N galaxies. These AGNs were first identified as a pervasive type among the more distant radio galaxies. We now recognize that many of their

11.9 The radio galaxy NGC 1068, which is also a Seyfert galaxy.

characteristics are quite similar to those of Seyfert galaxies; the N galaxies differ mostly by being farther away.

We have already mentioned another type of radio galaxy, the cD galaxy — the fat cannibals that emit radio radiation from recent gorging on their cluster neighbors (see Chapter 9). They do not look like N galaxies; quite the contrary, they are bloated and diffuse. But many of the same mechanisms are at work in them, as well as in a variety of other galactic oddities. The champions, though, among a terrific line-up of characters, are the quasars, the subject of the next chapter.

Quasars

12

Active galaxies are quite remarkable, but they hardly hold a candle to a group of mysterious, enigmatic objects that were first detected in the 1960s. Radio astronomers, already fully aware of radio galaxies as well as various kinds of local Milky Way sources, began to find something new — tiny, bright radio sources that were not identifiable as any of the familiar things (fig. 12.1). They appeared small, like stars, but seemed to lie in parts of the sky that were empty of any galaxies, supernova remnants, or HII regions. As their positions were determined more and more accurately, the puzzle intensified. These brilliant radio sources were found to coincide in position with faint stars.

The First Quasi-Stellar Object

Radio astronomers, using new and very refined techniques, were able to establish a highly precise position for one of these sources, called 3C48, by 1960. When they plotted this position on the Palomar Atlas of the sky, the only object at that position turned out to be a rather faint star with no obvious peculiarity.

The optical identification of 3C48 with a star almost seemed to be a step backward. Years before, in the infancy of radio astronomy, most radio sources were thought to be stars and were called radio stars. As more knowledge accumulated, most of them turned out to be other kinds of objects — gas clouds, supernova remnants, galaxies. Suddenly,

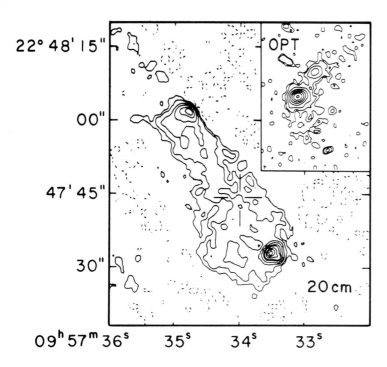

12.1 Maps of the radio radiation and the optical brightness (inset) for the quasar known as 0957+227.

in 1960, there seemed to be a genuine radio star. But what kind of star could it be? What weird happening could make a star radiate such immense amounts of radio energy? To answer this question, the world's largest telescope was turned to look at it, first directly and then with a spectrograph.

The first photograph centered on the radio position of 3C48, taken by the Palomar 200-inch telescope, was very exciting but also puzzling. Right at the center was the star, and as though to confirm that it was the real source, a faint wisp of light pointed to it like a finger. Clearly this must be the thing, but what is it? astronomers asked. A spectrogram was obtained, but this did not help very much. In fact, the spectrum confounded everyone who looked at it. Instead of a continuous band of light of all different colors, like that of a star, the spectrum consisted of a faint band that was punctuated by a series of bright emission lines — all in the wrong places. The chemical elements commonly found in stars and gas clouds have characteristic patterns of emission line wavelengths, and none of them seemed to correspond to the lines in the spectrum of 3C48. The mysterious new radio sources were star-like in appearance, but they seemed to be made of unrecognizable material.

Initially the discoverers called these objects "quasi-stellar radio sources." This cumbersome name was all right when there were only two or three known, but as more and more of them were found, a shorter name was clearly needed. The decision had to be made between the initials, QSO, and a word made up from fragments of the words in the name, *qua*si-*stell*ar *r*adio source, or *quasar*. Being catchy and exotic sounding, *quasar* soon became the popular choice, and it has since come to be used for all kinds of nonastronomical things, from televisions to aircraft lights.

The Puzzle Solved

The next year saw the solution of the spectrum puzzle. In 1961 the Palomar telescope was used to obtain a spectrum of a 13th magnitude quasar, 3C273, the brightest of the quasi-stellar radio sources (fig. 12.2). When the plate was first developed, it looked rather like that of 3C48; there was a faint continuum superimposed upon a series of bright emission lines. But a pattern in the lines was recognizable that had not been apparent in the other spectrum. The lines were not in the right place, but they were arranged in the right way, with the normal spacing and intensities. All, however, were greatly displaced toward the red side of the spectrum (fig. 12.3). These otherwise normal lines were redshifted so far that they were almost missed. The same lines had, in fact, been there in the spectrum of 3C48, too, but were redshifted

12.2 3C 273, the brightest and nearest quasar. Notice the faint "jet" of light that seems to emanate from the central (highly overexposed) quasar image.

even farther. In its case, the pattern was not evident because some of the familiar lines were redshifted right off of the spectrum!

Redshifts of objects due to the Doppler effect were not unfamiliar to astronomers, of course. The astounding thing about the quasars was that the recessional velocities indicated were so immense. The largest velocities for stars in our galaxy are about 400 km/sec (about 1 million miles per hour). The quasars, which looked like stars, had redshifts as large as 150,000 km/sec (about 325 million miles per hour). They could not be stars in our galaxy, then, because they would be moving so fast that they would soon escape from the Milky Way to go hurtling out through intergalactic space. So what could they be?

Many hypotheses were suggested, some of them preposterous and some merely weird. Most were based on the assumption that the redshift was caused by the Doppler effect and that the quasars must, therefore, be moving rapidly away from us. Other ideas about the cause were also explored, however, including the possibility that a gravitational redshift might be involved. Einstein had shown that in the presence of a very strong gravitational field the wavelength of light is lengthened in a way similar to the Doppler effect. This shifting is visible (just barely) in the spectrum of the sun because of its large mass and is quite noticeable in the spectra of white dwarf stars, whose huge densities cause a significant gravitational redshift in the light emitted from their surfaces. However, as time went on and quasars with larger and larger redshifts were discovered, the gravitational redshift model had to be abandoned, as no physical system could exist, for any length of time, with such an immense gravitational field as these redshifts implied.

In the more than 25 years since quasars were discovered, huge amounts of data have been collected, but very slow progress has been made in our understanding. We now have identified over 3,000 quasars and find redshifts ranging from a few tenths to as large as 3.8 (these figures represent a comparison of the shift in wavelength to the rest wavelength).

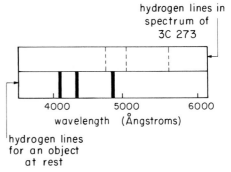

12.3 Quasar redshifts are like those of ordinary galaxies, only usually much larger.

Cosmological Controversy

The most likely explanation of the redshift settled down to be the so-called cosmological one: that quasars must be receding from us, like the galaxies, as part of the general expansion of the universe, and the enormous speeds indicated by the redshift would put them out among the most distant known galaxies. Some quasar redshifts are far greater than those of any measured galaxy—a fact that introduces yet another problem. The reason that no galaxies are known with such large redshifts is simply that galaxies so far away are too faint to be seen and measured. This implies that the quasars must be much brighter intrinsically than even the brightest galaxies. The luminosity of 3C273, for instance, is calculated to be about 100 times that of a normal giant galaxy. Yet it looks like just a star.

And if this problem were not enough, it has turned out that many of the quasars are variable in their brightness. For example, it was found that 3C273, which had been recorded on patrol plates taken at Harvard Observatory for over 50 years, had been varying in brightness irregularly over that interval of time. When astronomers measured the brightnesses of quasars with photoelectric photometers, they found that some were highly variable, sometimes by as much as a factor of 100. In a few cases, the change was rapid, with variations happening in a matter of only a day (fig. 12.4). This discovery introduced really serious problems into the cosmological interpretation of quasars. An object that changes so rapidly cannot be very big. Light travels one light-day in one day, so if an object can show a large overall change in its brightness in so short a time, it must be smaller than one light-day; otherwise, any changes would be smeared out by the time it takes for the light from the far side to reach the near side (fig. 12.5). A light-day is only about the

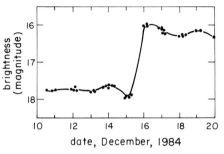

12.4 Variations in quasar luminosities tend to occur rapidly and irregularly.

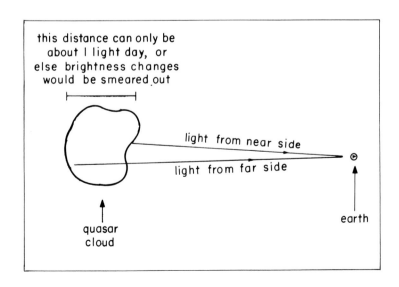

12.5 The size-vs-brightness-fluctuations argument shows that quasars must be small, or else their brightness changes would appear to be smeared out by the different times it would take light to travel to us from their different regions.

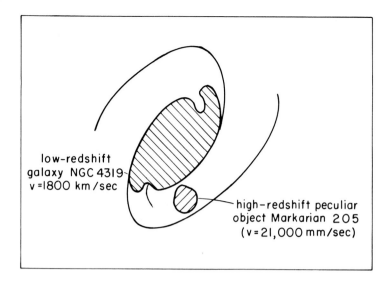

low-redshift
galaxy NGC 4319
v = 1800 km/sec

high-redshift peculiar
object Markarian 205
(v = 21,000 mm/sec)

12.6 One of the puzzling cases of a high-redshift peculiar object (in this case an emission-line galaxy known as Markarian 205) that appears to be associated with a low-redshift galaxy, NGC 4319.

size of the solar system (Pluto's orbit is actually about half a light-day in diameter). How could an object only the size of the solar system emit light 100 times more powerfully than a galaxy of hundreds of billions of stars? Nobody knew. Radio astronomers made things worse by using their newly developed intercontinental interferometry techniques to measure quasar diameters directly, also finding them to be very small and complex in structure.

The cosmological explanation seemed such an impossibility that many astronomers thought the quasars must somehow be local, after all. Perhaps they are stars shot out of our galaxy in some way at enormous speeds. Or maybe the redshift is due to some new physical phenomenon other than the Doppler effect. This last possibility seemed to be supported when astronomers found apparent connections and alignments between two or more objects with wildly different redshifts (fig. 12.6). One suggestion was that these objects demonstrate some unexpected effect of aging on light, whereby the wavelength of light changes with time. But this is a maverick idea; at present most astronomers believe that the examples of alignments and connections are merely optical illusions, and that the objects are in fact at vastly different distances.

The Controversy Resolved?

The 1980s brought new light to the study of the enigmatic quasars. Astronomical equipment had been developed at last

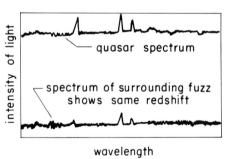

quasar spectrum

spectrum of surrounding fuzz
shows same redshift

intensity of light

wavelength

12.7 The spectrum of the "fuzz" around a quasar is found to have the same redshift as the quasar, indicating that the object is probably a distant galaxy with a disturbed superluminous quasar nucleus.

that could penetrate the space immediately surrounding the blazing center of a quasar. The question had been asked before but could not be answered: is there a galaxy there, hidden by the brilliant, overexposed image of the quasar? Are the quasars at the centers of galaxies? Are they somehow related to galaxies with active nuclei, such as the Seyfert galaxies? If only we could see a faint galaxy image right up close to the quasar, surrounding it, we would know that the answer is yes.

Some of the first successful tests were carried out for the quasar 3C273, which was found to have something surrounding it that seemed to have a light distribution like that of a normal elliptical galaxy. Spectra of this faint fuzz around quasars showed a redshift that is identical to the redshift of the quasar itself (fig. 12.7). Therefore, if the fuzz really is a galaxy, then it is receding with a velocity that corresponds to its great distance, and therefore the quasar, too, must be at the cosmological distance and must be partaking of the cosmic expansion. By 1983, through the studies of several enterprising groups who used CCD detectors to look at the properties of the light around quasars, it was finally established that they are, indeed, at the centers of galaxies (fig. 12.8). Many of the quasar galaxies turn out to be barred spirals and an unusually large percentage of them have smaller companions. We do not yet know why double galaxies should preferentially be quasars, or why there should be so many barred galaxies among them. But we finally know that quasars are terribly luminous, tiny, almost unbelievably energetic things that are going off at the centers of galaxies. The problem left is to find out why.

The Quasar Cloud

Even before they knew where quasars are, astronomers had built up a fairly detailed picture of the cloud of material that forms the object (fig. 12.9). The spectra of quasars give enough information for us to conclude that there is a small, compact object at the very center, which is surrounded by a number of very hot clouds of gas, as well as some regions of cooler gas. Dust clouds seem to be interspersed throughout the cloud, and they have high velocities, as if they have been ejected from the more central regions. There are high-energy particles all around, moving in a strong magnetic field and emitting synchrotron radiation. It is this radiation that makes some quasars such strong radio sources. Though it was the strong radio sources among them that initially led to

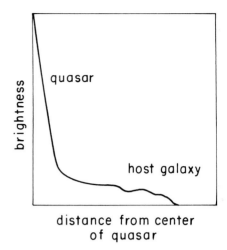

12.8 The brightness profile of a quasar is strongly peaked and the fainter light of the host galaxy surrounding it is difficult to detect.

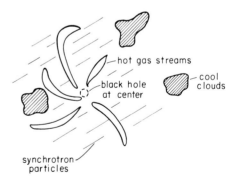

12.9 A model of the quasar cloud.

the discovery of quasars, most are radio-quiet. Consequently, there must be many variations in the physical properties of these remarkable objects.

Another instance of individual differences is the fact that some quasars do not show emission lines at all, indicating that they have no excited gas. With no lines in the spectrum, and hence no redshifts, it is not possible to learn the distance to these objects. These lineless quasars are called BL Lacertae objects, or blazars for short, after the prototype, which had been thought to be a variable star in our galaxy. When it was found that BL Lacertae is a strong radio source and when it turned out to have an optical continuum like that of a quasar, circumstantial evidence pointed to its identity as a quasar without emission lines. Now there are many other blazars known. We do not know the distances to most of them, except very roughly, judged from their brightnesses. For a few very important BL Lacertae objects, however, distances are known because it has been possible to find spectral lines in their surrounding fuzz. They turn out to be out among the distant galaxies and quasars, just as had been inferred from their several similarities to the quasars.

The Quasar Machine

One of the persistent difficulties posed by the quasars has been explaining the immense amount of energy they emit. Astronomers have tried many different things, including all the explanations put forward earlier to explain the radio galaxies (Chapter 11). Only one mechanism seems to be capable of producing so much energy: gravitational collapse of a massive object. Most astronomers believe that the centers of quasars, like the centers of active galaxies, consist of massive collapsed objects, such as black holes. There is no real proof that a black hole is involved; it is merely the only thing anyone has thought of that could do the trick. If a quasar consists of a black hole, then its mass would have to be about 100 million suns. A very bright quasar would have to be swallowing mass from the surrounding galaxy at the rate of about 100 solar masses per year.

In the standard quasar model, the black hole is surrounded by a spinning accretion disk. The power comes from the potential energy of the infalling gas. The disk supplies thermal energy to produce the light and probably involves magnetic flares, jets of lined-up particles, and other exotic phenomena. The high-energy gas particles form a plasma that can explain the x-rays emitted by some quasars.

Outside the environs of the accretion disk there is active gas that gives off the emission lines seen in quasar spectra. Farther out yet—probably completely outside the galaxy involved and merely seen by chance in front of it—is the cool gas that explains the absorption lines. This seems to come in two varieties: some of the absorption lines are arranged in the spectra in a way that suggests random clouds of hydrogen gas in the expanding space between us and the quasar. Other absorption lines are clustered at certain redshifts, suggesting hydrogen clouds in the outer parts of intervening galaxies in clusters of galaxies.

Gravitational Lenses

Quasars, though we may not know yet exactly what they are, are useful demonstrators of the validity of Einstein's general theory of relativity. Shortly after its publication, Sir Arthur Eddington predicted that the gravitational pull of massive objects should be able to focus light like a giant cosmic telescope, producing complex images of distant objects. Nearly 60 years later, in 1979, a gravitational lens was discovered in the form of a double quasar. The two components were only a few arc seconds apart, already a strange circumstance. But when the quasars' redshifts were measured, they were found to be identical. The only reasonable explanation was that we were looking at two images of the same quasar. Careful examination of the images revealed a faint object between them that had the properties of a galaxy. It is this galaxy, apparently, that acts as the lens, magnifying the image of the distant quasar and throwing two copies of it our way.

Four other examples were found in the following five years, all rather similar, with separations ranging from 2 to 7 seconds of arc (fig. 12.10). The lens system known as 2016+112 is a recent example. It consists of two images of a quasar that has a redshift of 3.27, placing it several billion light-years away. Next to these 23rd magnitude objects is an even fainter smudge that turns out to be an elliptical galaxy with a redshift of about 0.8, placing it considerably nearer than the quasar. This galaxy is the lens, bending the light of the vastly more distant quasar to form images that corroborate Eddington's prediction of long ago.

The Quasar Epoch

The redshifts of quasars bunch up at about 2 to 3, and hardly any quasars have been found with redshifts greater than 3.5.

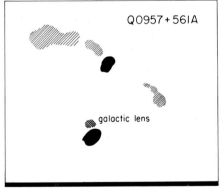

12.10 A double quasar image produced by a gravitational lens. The lensing galaxy is a nearer, giant elliptical galaxy, and it has formed two images of the far more distant quasar, known as 0957+561.

There are also very few quasars with small redshifts, so that the distribution of redshifts is bunched at a recessional velocity that corresponds to an early epoch in the history of the cosmos. Quasars were much more common then than they are now and than they had ever been before. This peculiar fact has led some astronomers to name this interval, representing only about 10% of the history of the universe, the quasar epoch. Why did quasars flare up back then? Why not earlier and why not now? These are some of the unanswered questions that still await some key insight or crucial observation.

Quasars are the most energetic and spectacular examples of the subjects of this book, the galaxies. The fact that so much still remains unknown about them and that there is still so much controversy about their basic nature is an indication of the marvels yet to be unraveled as we continue to explore the cosmos. The centuries have seen the expansion of our horizon from the earth to the stars to the galaxies. Now we stand at the edge of new and wonderous territory, the mysterious reaches of the quasars. And beyond them we can just glimpse the misty outlines of unimaginable glories — the beginnings of the universe and its magnificent bursting into bloom as an immense, beautiful, and hospitable world.

Sources of Illustrations

Index

Sources of Illustrations

2.5 A comparison of the spiral arm structure of NGC 3136 with the density wave theory predictions. (By permission of *The Astrophysical Journal.*)

2.6 Isophotes of the SBa galaxy NGC 1350. (By permission of *The Astrophysical Journal.*)

2.7 Isophotes of the SBb barred spiral galaxy NGC 1365. (By permission of *The Astrophysical Journal.*)

2.8 Isophotes of the irregular galaxy NGC 4449. (By permission of *The Astrophysical Journal.*)

4.1 Close-up photograph of M31 showing the central bulge of Population II stars and, on either side, Population I. (Photograph from Kitt Peak National Observatory.)

4.7 Photographic spectrum showing strong emission lines in the visual spectrum of the irregular galaxy NGC 5253. (Photograph from Cerro Tololo Interamerican Observatory.)

4.8 The distribution of HII regions in NGC 1232, a giant Sc spiral. (Photograph from Kitt Peak National Observatory.)

4.9 HII regions in the irregular galaxy NGC 4449, revealed by a hydrogen light photograph. (Photograph from Kitt Peak National Observatory.)

4.10 Dust lanes in the edge-on galaxy NGC 4565, silhouetted against its central bulge. (Photograph from Kitt Peak National Observatory.)

4.11 NGC 5194, a giant Sc spiral, and its small, distorted companion NGC 5195. (Photograph from Lick Observatory.)

5.1 A telescopic view of the stars in the Milky Way. (Photograph from Lick Observatory.)

5.6 The double galaxy NGC 5574-5576. (Photograph from Lick Observatory.)

5.7 The central part of the cluster of galaxies in Pegasus. (Photograph from Palomar Observatory.)

6.1 The Small Magellanic Cloud. (Photograph from Harvard College Observatory.)

6.2 The Large Magellanic Cloud. (Photograph from Harvard College Observatory.)

6.3 A portion of the Large Magellanic Cloud in which some of the many star clusters are identified. (Photograph from Cerro Tololo Interamerican Observatory.)

6.4 A close-up view of two neighboring clusters in the Large Magellanic Cloud, NGC 1844 (right) and NGC 1846 (left). (Photograph from Harvard College Observatory.)

6.5 NGC 1978, an old, highly elliptical cluster in the Large Magellanic Cloud. (Photograph from Cerro Tololo Interamerican Observatory.)

6.6 A comparison of the color–magnitude diagram for the LMC cluster NGC 2164 with a theoretical evolutionary path for a cluster 50 million years old. (By permission of *The Astrophysical Journal.*)

6.8 An area of the LMC that contains large numbers of supergiant stars. The cloud-like object is an HII region known as 30 Doradus. (Photograph from Harvard College Observatory.)

6.9 A section of the Large Magellanic Cloud with the Harvard variables identified by number. (Photograph from Harvard College Observatory.)

6.11 Light curves of six Large Magellanic Cloud Cepheids, measured in two colors. (By permission of Frances W. Wright and *The Astrophysical Journal.*)

6.12 The brightest and largest gas cloud in the Magellanic Clouds, known as the Tarantula Nebula, or 30 Doradus. (Photograph from Cerro Tololo Interamerican Observatory.)

6.13 The center of 30 Doradus, showing the embedded star cluster. (Photograph from Harvard College Observatory.)

6.14 The distribution of dust in the SMC from background galaxy counts. (By permission of *The Astrophysical Journal.*)

7.1 The large Sb spiral galaxy M81 in Ursa Major. (Photograph from Palomar Observatory.)

7.2 NGC 185, a companion to M31. (Photograph from Palomar Observatory.)

7.3 GR8, an extremely dwarf irregular galaxy in the local group. (Photograph from Lick Observatory.)

7.4 The stellar associations and star clusters of NGC 6822. (By permission of *The Astrophysical Journal.*)

7.5 The sizes of the stellar associations of NGC 6822, compared with those in other galaxies. (By permission of *The Astrophysical Journal.*)

7.6 Contours of equal star distribution in NGC 6822. (By permission of *The Astrophysical Journal.*)

7.7 IC 1613, a faint irregular galaxy in the local group. (Photograph from Lick Observatory.)

7.8 Three faint, young star clusters in IC 1613. (Photograph from Kitt Peak National Observatory.)

7.9 M32, a small but bright elliptical companion to M31. (Photograph from Kitt Peak National Observatory.)

7.10 Dust lanes in NGC 205. (Photograph from Palomar Observatory.)

7.11 Bright blue stars in NGC 205. (Photograph from Palomar Observatory.)

7.12 NGC 185, with its conspicuous dust lanes and bright blue stars. (Photograph from Palomar Observatory.)

7.13 NGC 147 isophotes, illustrating the galaxy's near perfect elliptical shape. (By permission of *The Astronomical Journal.*)

7.14 The Sculptor system. (Photograph from Harvard College Observatory.)

7.15 The Fornax galaxy, displaying its brightest stars and one of its globular clusters. (Photograph from Harvard College Observatory.)

7.16 The Leo I system. (Photograph from Kitt Peak National Observatory.)

7.17 The distribution of stars in Draco. (By permission of *The Astronomical Journal.*)

8.1 M31, the Andromeda galaxy. (Photograph from Palomar Observatory.)

8.2 Light curves of novae in M31, based on the data published by Hubble. (By permission of *The Astrophysical Journal*.)

8.3 Light curve of a Cepheid in M31 (Hubble's number 1, as identified in his discovery paper). (By permission of *The Astrophysical Journal*.)

8.4 Period–luminosity diagram for M31 Cepheids as determined by Baade and Swope for a small region in the extreme southwest portion of the object. (By permission of *The Astronomical Journal*.)

8.5 Two photographs of an outer portion of the Andromeda galaxy. The left-hand photo was taken with a filter that admits the red light of hydrogen gas (H-alpha) and the right-hand photo with a filter that excludes this light, registering only starlight. (Photographs from Kitt Peak National Observatory.)

8.7 Continuum radio map of M31 made with the Bonn 100-meter radio telescope. (From the Max Planck Institute for Radioastronomy and the European Southern Observatory.)

8.8 A small open star cluster in M31. (Photograph from Kitt Peak National Observatory.)

8.9 Stellar associations in M31. (Photograph from Kitt Peak National Observatory.)

8.10 A portion of M31 containing clusters and associations. (Photograph from Kitt Peak National Observatory.)

8.12 The dust arm of M31. (Photograph from Kitt Peak National Observatory.)

8.13 Comparison of the neutral hydrogen arms and the open cluster distribution in M31. (By permission of *The Astronomical Journal*.)

8.14 M33, an Sc galaxy in the local group. (Photograph from Palomar Observatory.)

8.15 HII regions in a portion of M33. (Photograph from Kitt Peak National Observatory.)

8.16 Spectrum of a supergiant star in M33 (star B324), compared with the spectra of two standard galactic supergiants of similar temperature and luminosity. (By permission of Roberta Humphreys.)

9.1 The Hercules cluster of galaxies. (Photograph from Kitt Peak National Observatory.)

9.2 The Coma cluster of galaxies. (Photograph from Palomar Observatory.)

9.3 X-ray map of the cluster of galaxies Abell 85, which is dominated by a giant cD galaxy in its center. (By permission of W. Forman and C. Jones.)

9.4 The central part of the Virgo cluster of galaxies. (Photograph from Palomar Observatory.)

9.5 Two spiral galaxies in the Virgo cluster. (Photograph from Lick Observatory.)

9.6 Three galaxies belonging to the NGC 5566 group, a small cluster. (Photograph from Lick Observatory.)

10.3 The largest HII regions in NGC 2403 show up as diffuse clouds in this photograph taken in hydrogen light. (Photograph from Kitt Peak National Observatory.)

10.4 HoII, a galaxy in the NGC 2403 group. (Photograph from Lick Observatory.)

10.6 Dust in the Milky Way shows up by the way it obscures more distant stars in photographs of star fields like this. (Photograph from Lick Observatory.)

10.8 A very distant cluster of galaxies is just barely visible in this deep photograph. (Photograph from Palomar Observatory.)

11.1 A radio telescope, such as this one of the Harvard-Smithsonian Center for Astrophysics, looks something like a television receiver, except that it can turn to any point in the sky. (Photograph from Harvard College Observatory.)

11.2 The five-kilometer radio interferometer of Cambridge University. (Photograph from Cambridge University.)

11.4 NGC 520, two galaxies that are probably merging, though they are not a radio source. (Photograph from Manastash Ridge Observatory.)

11.5 The arrow in this figure points to 3C 31, a radio galaxy in a small cluster of galaxies. (Photograph from Palomar Observatory.)

11.6 M82, a remarkable radio galaxy in a nearby group. (Photograph from Manastash Ridge Observatory.)

11.7 A radio map of the radio galaxy 3C 430, made with the VLA. (By permission of *The Astronomical Journal.*)

11.8 Map of the radio radiation from the nearby radio galaxy Fornax A. (By permission of *The Astronomical Journal.*)

11.9 The radio galaxy NGC 1068, which is also a Seyfert galaxy. (Photograph from Palomar Observatory.)

12.1 Maps of the radio radiation and the optical brightness (inset) for the quasar known as 0957+227. (By permission of *The Astronomical Journal.*)

12.2 3C 273, the brightest and nearest quasar. (Photograph from Kitt Peak National Observatory.)

Index